Glossary of Dairy Technology

GLOSSARY OF DAIRY TECHNOLOGY

Dr. Deepak Mudgil
Assistant Professor
Department of Dairy and Food Technology
MIDFT, Mehsana, Gujarat

&

Dr. Sheweta Barak Mudgil
Assistant Professor
Department of Dairy and Food Technology
MIDFT, Mehsana, Gujarat

SCIENTIFIC
PUBLISHERS (INDIA)

Published by:

Scientific Publishers (India)
5 A, New Pali Road, P.O. Box 91
Jodhpur 342 001 (India)

Branch Office
Scientific Publishers (India)
4806/24, Ansari Road, Daryaganj
New Delhi - 110 002 (India)

E-mail: info@scientificpub.com
Website: www.scientificpub.com

ISBN: 978-81-7233-928-9

Typeset by Rajesh Ojha
Printed in India

Dedicated

to

Maa Saraswati & our family

CSIR – Central Food Technological Research Institute, Mysore – 570020
(Council of Scientific and Industrial Research, Ministry of Science & Technology, Govt. of India)
सीएसआईआर - केंद्रीय खाद्य प्रौद्योगिकी अनुसंधान संस्थान, मैसूर - ५७० ०२०
(वैज्ञानिक तथा औद्योगिक अनुसंधान परिषद्, विज्ञान और प्रौद्योगिकी मंत्रालय, भारत सरकार)

Prof. Ram Rajasekharan, FNA, FASc, FNASc, FNAAS
Director
प्रो. राम राजसेखरन, एफएनए, एफएएससि, एफएनएएससि, एफएनएएएस
निदेशक

MESSAGE

I was delighted when Dr. Deepak Mudgil approached me to write a message for the book, "Glossary of terms: Diary Technology" written by Dr. Deepak Mudgil & Dr. Sheweta Barak, because this is a valuable asset for persons who wish to learn about diary sector and also pursue a career in dairy and related fields. Apart from serving the dairy industries, it will serve as a handbook for all students in the field of food science and technology. As the Director of CSIR-CFTRI, I am of the opinion that this book would be a readable asset in Libraries of food science institutes and academic institutions of our country. Although in today's information technology driven world, it will be easy to quickly "google" the meaning of an unknown word, it is quite important to have such one-stop books getting published regularly. The contents of the book are impressive and very exhaustive. I congratulate the authors for their efforts in coming out with such a book for the benefit of students and diary entrepreneurs, and I wish that it sees a lot of beneficiaries in the country.

Wishing the book all success and wide readership,

With best regards,

Prof. Ram Rajasekharan
Director, CSIR-CFTRI,
Mysore-570020

An ISO 9001:2008 & ISO 14001:2004 Organization www.cftri.com An NABL Accredited Laboratory
Phone: +91 821 2517760 | Fax: +91 821 2516308 | E-mail: director@cftri.res.in; ram@cftri.com

MIDFT

AN INITIATIVE OF DUDHSAGAR DAIRY

विद्या • कौशलम् • अभिवृत्ति

MANSINHBHAI INSTITUTE OF DAIRY & FOOD TECHNOLOGY

Foreword

The urbanization & changes in life style of Indian population has given new dimension to food habits & food consumption pattern. This has generated demand of qualified professionals in Dairy & Food Technology.

This book gives information on important terms commonly used in DAIRY SCIENCE/ DAIRY TECHNOLOGY to students as well as working professionals in dairy plants. It will be a good handy reference book for students & dairy professionals during their studies, at workplace and in future career.

The authors have compiled good information on the subject and covered all important terms related to dairy science and technology.

I wish that the dairy fraternity make full use of the book.

Pramthesh R Patel
Executive Director,
MIDFT, Dudhsagar Dairy,
Mehsana

PREFACE

Dairy & food technology is coming as an emerging field for students from career point of view. In India, many national level institutes and universities are offering courses related to dairy and food technology at graduation, post graduation and doctorate level. Admission in these institutes at different levels is carried out by various types of entrance exams which are conducted every year. ICAR conducts ASRB-NET exam in Dairy & Food Technology disciplines which is an essential requirement for the eligibility to become assistant professor/ lecturer in colleges or universities. ICAR also conducts JRF-SRF exams every year for students who want to pursue research in agriculture related fields. Indian Institute of Technology (IIT) conducts GATE exam annually for admission to M. Tech. courses in IITs. This book will be helpful for the student aspirants preparing for such exams and interviews. This book contains about 1000 technical and general terms frequently used in the dairy sector. The terms in the book cover the topics related to market milk, dairy processing, fat rich dairy products, cheese and fermented milks technology, traditional dairy and food products, ice cream and frozen desserts, condensed and dried milk, by-products technology and packaging technology. This book will be helpful for students as well as dairy professionals working in dairy & food industries, for preparation of viva voce examination and job interviews, respectively. We would like to show our sincere thanks to Dr. K.G. Upadyay (Retd. Professor, Dairy Technology), who gave us the idea for writing this book. We are also thankful to Sh. P.R. Patel (Executive Director, MIDFT, Mehsana) for providing facilities and motivating us for writing this book. We wish to acknowledge authors and editors of all the books and research publications which helped us in the preparation of this book. The patience of Scientific Publishers (India) and the assistance provided by Scientific Publishers staff members are highly appreciated. Last but not least we must thank our family for their love, support and encouragement.

Dr. Deepak Mudgil
Dr. Sheweta Barak Mudgil

A

Accelerated ripening: It is the process of reducing the ripening time of cheese to lower the overall cost of cheese storage by increasing the ripening temperature, use of enzyme, increasing the numbers of starter bacteria and increasing the rate of lysis of starter cells. Accelerated ripening processes are primarily aimed at accelerated flavor formation (especially a stronger protein break-down), while maintaining a satisfactory texture.

Acceptable Daily Intake (ADI)**:** It is a measurement for the products safeness. It is given in mg/ kg body weight, meaning a person should multiply the ADI with one's body weight to get the maximal dosage without causing health problems. The ADI is not set as of giving an immediately effect, but as a daily dosage over a person's normal living time. In most cases the average dosage over longer time is more important than the dosage from day to day.

Accreditation: It is a procedure by which an authoritative body gives formal recognition that a body or a person is competent to carry out specific tasks.

Accretion: A gradual process in which layers of a material are formed as small amounts are added over time.

Accuracy: The closeness of the obtained analyte value to the true value; it can be established by analysing a suitable reference material.

Acetoin: This is the precursor of diacetyl, the aroma substance of cream and butter.

Acid casein: The casein precipitated from skim milk by various acids such as hydrochloric acid, sulphuric acid, nitric acid or lactic acid is known as acid casein. Acid casein is insoluble in water and poorly dispersible in alkaline solution.

Acid coagulation: It is the process of milk clotting by lowering the pH. This clotting or coagulation is due to the precipitation of casein at its isoelectric pH.

Acid flavor (cheese defect): The cheese is sharp and puckery to the taste or has a taste which is characteristic of lactic acid.

Acid whey: Acid whey, also known as "sour whey" is a by-product produced during the making of acid types of dairy products such as cottage cheese or strained yogurt.

Acid-cut (cheese defect): Bleached or faded appearance which sometimes varies throughout the cheese but is present most often around mechanical openings.

Acidophilus milk: It is fermented milk prepared by lactic acid fermentation by *Lactobacillus acidophilus*. Acidophilus milk has some health benefits due to probiotic behaviour of *Lactobacillus acidophilus*.

Acidowhey: Acidowhey is a health beneficial fermented beverage produced from whey, the residual liquid by-product separated out from milk during the preparation of paneer and cheese.

Acidulants: Acidulants are acids used in processed foods for a variety of functions that enhance the food. Acids are used as flavoring agents, preservatives in microbial control, chelating agents, buffers, gelling and coagulating agents, and in many other ways.

Active packaging: Active packaging refers to the incorporation of certain additives into packaging film or within packaging containers with the aim of maintaining and extending product shelf life. Active packaging includes additives or freshness enhancers that are capable of scavenging oxygen; adsorbing carbon dioxide, moisture, ethylene and/or flavour/odour taints; releasing ethanol, sorbates, antioxidants and/or other preservatives; and/or maintaining temperature control.

Additives: These are the substances or mixture of substances other than basic food ingredients added to food products for specific functions.

Adsorption: Some molecules in a solution that is in contact with a phase surface can accumulate at this surface, forming a monolayer. This is called adsorption. A substance that does adsorb is called a surfactant.

Adulterant: An adulterant is a chemical which acts as a contaminant when combined with other substances. Adulterants are added to pure substances to extend the quantity while reducing the quality.

Aflatoxins: Aflatoxins are naturally occurring mycotoxins that are produced by *Aspergillus flavus* and *Aspergillus parasiticus*, species of fungi. Aflatoxins are toxic and among the most carcinogenic substances known.

Aftertaste: The last flavor sensation perceived after tasting a food product.

Ageing: Also known as curing or ripening, aging is the process of holding cheeses in carefully controlled environments to allow the development of microorganisms that usually accentuate the basic cheese flavors.

Agglomeration: The production of granules from powder particles

Agglutinin: An agglutinin is a substance that causes particles to coagulate to form a thickened mass. Agglutinins can be antibodies that cause antigens to aggregate by binding to the antigen-binding sites of antibodies or they can be any substance other than antibodies such as sugar-binding protein lectins. Agglutinins work by clumping on particles causing the particles to change from fluid-like state to thickened-mass state.

Aggregation: Aggregation refers to the state in which particles remains together for a longer time than they remain in the absence of the attractive forces between them. Example: flocculation is reversible type of aggregation whereas coagulation is reversible type of aggregation.

Agmark: Agricultural Produce (Grading and Marking) Act 1937, which prescribes grade standards for agricultural and allied commodities. These are known as "Agmark" standards. Grading under the provision of this act is voluntary. Manufacturers who comply with standard laid down by Directorate of Marketing and Inspection (DMI) are allowed to use "Agmark" labels on their products.

Air cells: These are the structures in dairy products such as whipped cream, butter and ice cream which enclose air in it and have contribution in the texture and consistency of the product.

Alcaligenes: Alcaligenes is a genus of gram-negative, aerobic, rod-shaped bacteria. The species are motile with one or more peritrichous flagella. Alcaligenes species have been used for the industrial production of non-standard amino acids.

Alfa process/ Alfa-Laval process: This is the process of continuous butter making process. This was developed simultaneously in both Germany and Sweden. Subsequently the German process retained the name of Alfa while the Swedish one was changed to Alfa-Laval.

Algorithms: Software building blocks used to construct control sequences in computerised process control.

Alkaline flavor: It is a type of flavor defect in cream which may occur due to over-neutralization of cream using a neutralizer.

Alkaline phosphatase: An enzyme in raw milk having a similar *D*-value to heat-resistant pathogens, used to test for effectiveness of pasteurisation.

Allergens: Food products that can trigger severe allergic reactions. Theoretically almost any food may be implicated, but the most common are peanuts and tree nuts.

Allergy: An allergy is a hypersensitivity disorder of the immune system.

Amino acid: Amino acids are organic compounds composed of amine (-NH_2) and carboxylic acid (-COOH) functional groups, along with a side-chain specific to each amino acid that combine to form proteins. The key elements of an amino acid are carbon, hydrogen, oxygen, and nitrogen, though other elements are found in the side-chains of certain amino acids. Amino acids and proteins are the building blocks of life. When proteins are digested or broken down, amino acids are left.

Ammix process: In Ammix process, fresh milk fat is mixed with cream and salt and shock cooled to give rapid crystallisation.

Amul: Brand name (Anand Milk Union Limited) for products of Gujarat Co-operative Milk Marketing Federation (GCMMF).

Anhydrous Milk Fat (AMF): Anhydrous milk fat is a product made from fresh cream or butter by removal of practically all of the moisture and non-fat solids. It is a product containing not less than 99.9% fat and 0.25% sterol (w/w), 0.15% free fatty acids (w/w) and free from additives. Water content of AMF should not exceed 0.1%, if water content is higher upto 0.4%, the product is generally designated as "butter oil".

Annatto color: The pigment bixin found in the coating of the annatto seed. The color hue ranges from yellow to reddish-orange. It is used in dairy products such as butter, cheese etc. Annatto is odorless, tasteless and is not a preservative.

Antibiotic: It is a drug used to destroy or slow down the growth of bacteria. It is used to treat the infections caused by bacteria.

Antibodies: Glycoprotein molecules produced in the body in response to the introduction of an antigen.

Anticaking agent: These are the substances which prevent a powder to cake e.g. when a product is sticky or is in contact with moisture. Anticaking agents are generally used in powdered food such as milk powder, salt, powder soups etc.

Antigen: Any substance which promotes the formation of antibodies in the blood of an animal, e.g. bacteria, bacterial toxin, some proteins etc.

Antioxidants: These are the substances that prevent oxidation of food. Fat containing products can get rancid when the fat in a product gets in contact with oxygen in the air. Antioxidants can be added to prevent the products going bad due to oxidation. Antioxidants can be vitamins as e.g. vitamin E, spices and natural extract as well as syntactically produced.

Antitoxin: Antibody to a toxin capable of reacting with that poison and neutralizing the toxin.

AOCS: American Oil Chemists' Society.

Aroma: It is the property of a material which affects the olfactory sense. Flavor consists of taste and aroma.

Aseptic canning: A system of canning in which sterilized foods are canned under aseptic conditions e.g. in an atmosphere of live steam or in an atmosphere sterilized by ultra-violet radiation.

Aseptic packaging: Aseptic packaging can be defined as the filling of a commercially sterile product into a sterile container under aseptic conditions and hermetically sealing the containers so that re-infection is prevented. This results in a product, which is shelf-stable at ambient conditions. The term "aseptic" is derived from the Greek word "septicos" which means the absence of putrefactive micro-organisms.

Aseptic: Free from living micro-organisms, spores and viruses.

Astringent: A term descriptive of a harsh taste.

Atomisation degree: Producing droplets of specific size and surface area by atomization is the most critical step in the spray drying process. The degree of atomization, under a set of spray drying conditions, controls the drying rate, and therefore the required particle residence time, and therefore the spray dryer size.

Atomiser: A device to form fine droplets of liquid (e.g. in a spray drier). It may be a high pressure nozzle or a perforated spinning disk through which the liquid is pumped.

Atomization: It is the process of breaking concentrate into fine droplets to provide large surface from which evaporation can take place in the spray drier. It can be achieved by disc atomizer or nozzle atomizer.

Auto oxidation: The oxidation of a substance upon its exposure to air.

Autoclaving: Autoclaving is a process used to destroy micro-organisms and decontaminate microbiological instruments by using high pressure and high temperature steam (121°C) for sterilization. The device used for autoclaving is called autoclave.

Automatic inoculation system (AISY): AISY in starter culture technology was patented by Chr. Hansen in Denmark in 1990s in collaboration with Tetra Pak in Sweden. It uses DVS concentrated starter cultures by integrating and automating the thawing or rehydration, mixing and dosing operation from the starter culture pack right through to the fermented milks/cheese vats/tanks.

Automation: Automation is the use of various control systems for operating equipment such as machinery, processes in factories, boilers and heat treating ovens etc. The biggest benefit of automation is that it saves labor; however, it is also used to save energy and materials and to improve quality, accuracy and precision.

B

Bacterial thickening: This is a type of defect in sweetened condensed milk caused by bacteria such as *Streptococcus albus* and *Micrococcus freudenreichii*. These organisms increase the viscosity of milk via fermentation and coagulation.

Bacteriocins: Bacteriocins are proteinaceous toxins produced by bacteria to inhibit the growth of similar or closely related bacterial strain(s). They are phenomenologically analogous to yeast and paramecium killing factors, and are structurally, functionally, and ecologically diverse.

Bactofugation: Bactofugation is a process in which a specially designed centrifuge called a Bactofuge is used to separate bacterial cells and spores from milk. Bactofugation is also used to improve the bacteriological quality of milk intended for other products like cheese, milk powder and whey for baby food. Bactofugation removes 95% of the spores of milk.

Bactofuge: Bactofuge are special nozzle clarifying separator with high separation precision that can remove micro-organisms from milk based on their density difference (skim milk – 1.036; bacteria – 1.07 – 1.13 g/cm^3).

Bag filtration: It is a pressure driven operation, wherein the product to be filtered is forced through the filter unit which consists of filter housing, a retaining basket and the filter bag.

Baker's cheese: A pasteurized cheese produced from skimmed cow milk. It is similar to cream cheese but contains virtually no fat. Although it is less rich, it is used in many pastry products because it is also less expensive.

Bal mithai: It is a heat desiccated product prepared from khoa. It is chocolate brown in color and topped with small sugar granules, giving the surface a distinct mosaic look. It has a rubbery body and chewy texture.

Balance tank: The balance tank provides a constant supply of milk. It is equipped with a float valve assembly which controls the liquid level nearly constant ensuring uniform head pressure on the product leaving the tank. The balance tank also prevents air from entering the pasteurizer. The balance tank provides a means for recirculation of diverted or pasteurized milk.

Bar coding: A barcode is an optical machine-readable representation of data relating to the object to which it is attached. Later they evolved into rectangles, dots, hexagons and other geometric patterns in two dimensions (2D).

Barny flavor: This term is used by flavor panelists to describe the milk flavor associated with a poorly ventilated barn. Primary cause of barny flavor is inadequate ventilation.

Basundi: Basundi is an Indian dessert mostly in Maharashtra, Gujarat and Karnataka. It is sweetened dense milk made by boiling milk on low heat until the milk is reduced by half. Heavy cream may be added during the boiling process to hasten the thickening process. Once reduced, a little sugar, cardamom, charoli and/or saffron are added. Basundi is served chilled, often garnished with slices of almonds and pistachios.

Baudouin's test: This is a test for detection of adulteration of butter by sesame oil. The fat is melted and shaken with a solution of cane sugar in hydrochloric acid. If the acid layer goes pink it indicates the presence of the oil.

Baume hydrometer: The Baume hydrometer, named for the French chemist Antoine Baume, is calibrated to measure specific gravity on evenly spaced scales; one scale is for liquids heavier than water, and the other is for liquids lighter than water.

Bavarian blue mould cheese: It is a soft cheese which has been matured through the mould fungus "*Penicillium roqueforti*" whereby the fungus develops inside the cheese and not on the surface.

Bend testing: Bend testing determines the ductility or the strength of a material by bending the material over a given radius. Following the bend, the sample is inspected for cracks on the outer surface. Bend testing provides insight into the modulus of elasticity and the bending strength or a material.

Beta-carotene: It is a type of carotenoid which is responsible for the yellow color of milk fat

Binders: A binder is any material or substance that holds or draws other materials together to form a cohesive whole.

Biofilm: A biofilm is an aggregate of microorganisms in which cells that are frequently embedded within a self-produced matrix of extracellular polymeric substance (EPS) adhere to each other and/or to a surface.

Biogas: Biogas typically refers to a gas produced by the breakdown of organic matter in the absence of oxygen. It is a renewable energy source, like solar and wind energy. Biogas is produced by anaerobic digestion with anaerobic bacteria or fermentation of biodegradable materials such as manure, sewage, municipal waste, green waste, plant material, and crops.[1] It is primarily methane (CH_4) and carbon dioxide (CO_2) and may have small amounts of hydrogen sulphide (H_2S), moisture and siloxanes.

Bioghurt (trade name): It is a commercial trade name of probiotic fermented milk made with various lactic acid bacteria including Bifidobacteria in addition to *S. thermophilus* and *Lb. delbrueckii* ssp. *bulgaricus*.

Biological value (BV): The Biological Value (BV) of a Protein is a value that measures how well the body can absorb and utilize a protein. The higher the Biological Value of the protein you use, the more nitrogen your body can absorb, use, and retain. As a result, proteins with the highest BV promote the most lean muscle gains. Whey protein has the highest BV value, rating as a 104. Egg protein is only second to whey rating as a 100 with milk proteins being a close third rating as 91. Beef rates as an 80 with soy proteins a distant 74. Bean proteins, due to the fact that are plant-based proteins, only rate a 49.

Biosynthesis (milk): Milk is synthesized in the mammary gland. Within the mammary gland is the milk producing unit, the alveolus. It contains a single layer of epithelial secretory cells surrounding a central storage area called the lumen, which is connected to a duct system. The secretory cells are, in turn, surrounded by a layer of myoepithelial cells and blood capillaries. The raw materials for milk production are transported via the bloodstream to the secretory cells. It takes 400-800 L of blood to deliver components for 1 L of milk.

BIS: Bureau of Indian Standards is a National Standards Body which is successfully promoting and nurturing standards movement within the country since 1947. BIS is involved in standards formulation, certification, foreign manufacturers scheme, registration scheme and testing and calibration services.

Bitter flavor (cheese defect): A distasteful flavour characteristic of quinine which is most frequently found in aged cheese varieties.

Bitterness (protein hydrolysates): Bitterness is a natural consequence of development of bitter-tasting components during enzymatic hydrolysis of protein. It is the main defect that limits the use of protein hydrolysates for human consumption.

Bitty cream: Bitty cream phenomenon is produced by the lecithinase activity of *Bacillus cereus*. This enzyme hydrolyses the phospholipids associated with the milk fat globule membrane to produce small proteinaceous fat particles which float on the surface of hot drinks and adhere to surfaces of crockery and glasses.

Biuret test: A chemical test for protein.

Black spot in butter: It is a defect in butter which may be caused by bacteria or moulds.

Blast freezing: It is quick freezing in which small ice crystals are formed and there is less damage to the food product.

Bleached surface (cheese defect): Faded colour begins at the surface of the cheese and progresses inward.

Blended spreads: These are the mixed type of spreads made from milk fat and vegetable/animal fat by blending in suitable proportions.

Blending: The mixing of two or more kinds of a product to attain a standardized product e.g. standardized milk and butter.

Bloats: Brown cans of evaporated milk and canned cream due to gas production by bacteria.

Bloom: Blooms are the small white spots on the surface of chocolate and candy which appears due to change in crystallization during storage. Temperature fluctuations generally speed up the bloom formation. Emulsifiers can be added to delay the formation of bloom, making the surface looks shiny and appetizing for longer.

Blowing: The bulging or bursting of tins of condensed milk usually due to the production of carbon dioxide by fermentation of the sugar by micro-organisms, usually yeasts.

Blue cheese: Blue cheese is a general classification of cow's milk, sheep's milk, or goat's milk cheeses that have had cultures of the mold *Penicillium* added so that the final product is spotted or veined throughout with blue, blue-gray or blue-green mold, and carries a distinct smell, either from that or various specially cultivated bacteria. Blue cheeses are typically aged in a temperature-controlled environment such as a cave. Blue cheese can be eaten by itself or can be crumbled or melted into or over foods.

BOD: Biochemical oxygen demand or B.O.D is the amount of dissolved oxygen needed by aerobic biological organisms in a body of water to break down organic material present in a given water sample at certain temperature over a specific time period. The BOD value is most commonly expressed in milligrams of oxygen consumed per litre of sample during 5 days of incubation at 20°C.

Body: It is a complex quality which denotes firmness and considers mainly viscosity and elasticity. It is used in grading butter and cheese.

Boiling Point Elevation: One of the colligative properties. The boiling point of a solution is increased over that of water by the presence of solutes, and the extent of the increase is a function of both concentration and molecular weight.

Booster Pump: It is centrifugal "stuffing" pump which supplies raw milk to the raw regenerator for the balance tank. It must be used in conjunction with pressure differential controlling device and shall operate only when timing pump is operating, proper pressures are achieved in regenerator, and system is in forward flow.

BOPP films: BOPP films are bi-axially oriented polypropylene films designed for flexible packaging.

Bound moisture: Liquid physically or chemically bound to a solid food matrix which exerts a lower vapour pressure than pure liquid at the same temperature.

Bovine Serum Albumin: Bovine serum albumin (BSA or "Fraction V") is a globular protein derived from cows (~66,000 Da) that is often used as a protein concentration standard in numerous biochemical applications due to its stability and lack of interference within biological reactions.

Bregott: It is a type of dairy spread developed and marketed by the Swedish Dairies Association (SMR). Bregott was originally developed using ripened cream (about 35% fat, pH 4.6-4.7) to which refined, deodorized soy bean oil was added in the batch butter churn. The soy bean oil was added at approximately 20% of the total fat.

Brine: Brine is a solution of salt (usually sodium chloride) in water. In different contexts, brine may refer to salt solutions ranging from about 3.5% (a typical concentration of seawater, or the lower end of solutions used for brining foods) up to about 26% (a typical saturated solution, depending on temperature).

Brining (cheese): In this process, the cheese is kept immersed in a concentrated solution of NaCl (brine) until the desired amount of salt has been absorbed.

Brownian motion: The random movement of microscopic particles suspended in a liquid or gas, caused by collisions with molecules of the surrounding medium.

Browning: Browning of milk occurs when milk is subjected to high temperatures to ensure microbiological safety such as pasteurization and sterilization. This browning occurs due to maillard reaction (non-enzymatic browning) which occurs when lactose reacts with the free amino acid side chains of milk proteins (mainly amino group of lysine residue) resulting in the formation of melanoidins.

Buffer: A solution that tends to resist the change in pH when acid or alkali is added.

Bulgarian Buttermilk: Bulgarian buttermilk is a version of cultured buttermilk in which the cream cultures are supplemented or replaced by yogurt cultures and fermented at higher temperatures for higher acidity. It can be more tart and thicker than cultured buttermilk.

Bulk Milk: It is raw milk from a dairy farm stored in a bulk tank; in contrast with packaged milk. Other bulk products, such as condensed skim milk and cream, may also be transported in bulk form.

Bulk modulus: It is the ratio of stress to the change in volume.

Bulk starter: Starter culture used in production is known as bulk culture.

Bulking agent: Bulking agents such as starch are additives that increase the bulk of a food without affecting its taste.

Bura: Bura is sugar heated so that it loses its water of crystallization and then powdered to get rid of any lumps which are formed.

Burfi: Barfi is a sweet confectionery made from condensed milk or khoa. Plain burfi is made with condensed milk and sugar cooked until it solidifies. Examples include besan barfi, Kaaju Barfi and Pista Barfi.

Butter fat: Also known as milk fat and is natural fatty portion or constituent of milk. Ghee is the name for anhydrous butter fat in India.

Butter grains: In butter manufacture, during churning partly destabilized globules adhering to the air bubbles collect in clusters cementing together by free fat. These clusters are known as butter grains.

Butter homogenizer: The machine which reduces the time it takes to defrost butter and improves the quality of the butter by creating the finest water dispersion and improves the spreadability of the butter.

Butter oil: It is a fat-rich dairy product made from fresh cream or butter by removal of moisture and non-fat solids. It has higher moisture content (0.4%) than anhydrous milk fat (0.1%).

Butter: Butter is a water-in-oil emulsion resulting from inversion of the cream which is an oil-in-water emulsion where the milk proteins act as emulsifiers. It may also be defined as a fat concentrate, obtained by churning cream, gathering the fat into compact mass and then working it.

Buttermilk concentrate: Buttermilk, the by-product of butter-making, is concentrated for use significantly in the manufacture of margarines and yellow fat spreads.

Buttermilk powder: It is also known as dry buttermilk and is obtained by removing water from liquid buttermilk that was derived from the churning of cream into butter and pasteurized prior to condensing. It contains about less than 5% moisture and about 4.5% or more milk fat by weight.

Buttermilk: The traditional buttermilk is the by-product obtained after churning butter out of cream. The different type of buttermilk includes cultured buttermilk, acidified buttermilk and powdered buttermilk.

Butyric acid: Butyric acid, also known under the systematic name butanoic acid, is a carboxylic acid with the structural formula $CH_3CH_2CH_2$-COOH. Butyric acid is found in milk, especially goat, sheep and buffalo milk, butter, Parmesan cheese, and as a product of anaerobic fermentation.

By-products: It is a general term applied to all materials other than desired product in any manufacturing process. It may also be defined as a product of commercial value produced during the manufacture of a main product. Dairy industry by-products include: whey, buttermilk, skimmed milk etc.

C

C/F ratio: Casein to fat ratio is an important parameter to be standardized in cheese milk for maintaining cheese quality. When this ratio is out of balance, the body of cheese is either too soft or too hard. For cheddar cheese, it is 0.7:1.

Caking: It is generally a deterioration or defect in powdered products which occurs when the water content becomes very high and the storage temperature is high which leads to the formation of cakes or lumps in powdered products.

Calandria: Heat exchanger used in an evaporator.

Calcium caseinate: Calcium caseinate finds use in both pharmaceutical preparations and as a food ingredient. The specifications for this product vary with its end use, but they frequently include a limitation of the calcium content to within the range of 1.0-1.5%. It forms micelles in water, producing an intensely white, opaque, 'milky' solution of relatively low viscosity. Calcium caseinates are much less soluble and have poorer functional attributes than sodium caseinate.

Canning: It is method of food preservation in which food products are placed in metal containers that are heated, exhausted and hermetically sealed.

Caramel: It is a dark-brown confectionery product made by heating sugars. It is used as flavoring and coloring agent in confectioneries.

Caramelization: It is the process of heating sugar slowly to around 170°C. As the sugar heats, the molecules break down and re-form into compounds with a characteristic color and flavor known as "caramel".

Caramelized milk concentrate: Caramelized milk concentrate or syrups are made by blending sweetened condensed sweet milk with reducing sugars and fats to give products which are used principally, in sugar and flour confectionery applications.

Carbery process: It is the best –known commercial operation of Express Dairies for the production of alcohol from whey streams.

Carbon dioxide: It is responsible for the eye (spherical holes) formation in Swiss cheese produced by lactic acid bacteria.

Carbonated yoghurt drinks: These are the beverages made from fresh yoghurt, water and spices. After mixing the basic ingredients the product is subject to carbonation and bottling. Examples include Doogh, Ayran etc.

Carbonation: The treatment of liquid with carbon dioxide especially under pressure. It usually improves taste and aroma, lowers the pH slightly and has a slight effect on keeping quality.

Carcinogen: Any substance that produces or promotes cancer.

Carotene: It is a class of unsaturated hydrocarbon substances which causes yellow color in milk fat. It is a orange photosynthetic pigment important for photosynthesis.

Carotenoids: Carotenoids are organic pigments found in the chloroplasts and chromoplasts of plants and some other photosynthetic organisms, including some bacteria and some fungi. These can be classified into two classes, xanthophylls and carotenes.

Carrageenans: These are the family of linear sulfated polysaccharides that are extracted from red edible seaweeds. There are three main varieties of carrageenan, which differ in their degree of sulfation. Kappa-carrageenan has one sulfate per disaccharide. Iota-carrageenan has two sulfates per disaccharide. Lambda carrageenan has three sulfates per disaccharide. They are widely used in the food industry due to their gelling, thickening and stabilizing properties.

Cartons: These are usually made of cardboard impregnated with paraffin wax and are probably the ideal container for milk as they are very light and virtually sterile because of their mode of manufacture and non-returnable.

Case hardening: It is the condition of the surface of solid or semi-solid food products in which shell at the surface of the particle solidifies and prevents further evaporation. It is found in various cheese varieties.

Casein hydration: It is generally considered to be the water that is more or less immobilized by a protein. Salt, pH and calcium level has a large effect on the extent of *para*-casein hydration.

Casein micelles: It consists of water, protein and salts. The protein consists of casein & salts in the micelles occur as a calcium, phosphate and citrates.

Casein: It is a most abundant protein in milk. It is relatively insoluble and tends to form structures called micelles that increase solubility in water. It consists of four types of polypeptides α_{s1}-casein, α_{s2}-casein, β-casein and κ-casein. They have some specific properties.

Caseinates: These are the by-products of dairy industry produced from acid casein curd or from dry acid casein by reaction under aqueous conditions with any one of the several different dilute alkalis. Examples include sodium caseinate and calcium caseinate.

Caseinomacropeptide: It is the soluble hydrophilic portion glycopeptide portion of kappa-casein obtained after ingestion or by action of rennet enzymes. It is responsible for increased efficiency of digestion, prevention of hypersensitivity to ingested proteins and inhibition of gastric pathogens.

Casson equation: This equation is used as an international standard for measuring the viscosity of food products. It also describes the flow behaviour of liquid and semi-solid foods.

Catalase: It is the enzyme (EC No. 1.11.1.6) which catalyses the decomposition of hydrogen peroxide to water and oxygen. It is found in milk leukocytes. Its optimum pH and temperature are 7 and 37°C, respectively.

Cathepsins: These are the proteases found in all animals. There are dozen members of this family, which are distinguished by their structure, their catalytic mechanism and which protein they cleave.

Caustic soda: It is sodium hydroxide and the caustic constituent of bottle-washing detergents also known as lye.

Cavitation: Cavitation is the formation of gas bubbles of a flowing liquid in a region where the pressure of the liquid falls below its vapor pressure. Cavitation is usually divided into two classes of behavior: inertial (or transient) cavitation, and non-inertial cavitation. Inertial cavitation is the process where a void or bubble in a liquid rapidly collapses, producing a shock wave. Such cavitation often occurs in control valves, pumps, propellers, impellers. Non-inertial cavitation is the process in which a bubble in a fluid is forced to oscillate in size or shape due to some form of energy input, such as an acoustic field.

Cellulose acetate membranes: Cellulose acetate membranes are extremely low binding filter membrane with pore sizes ranging from 0.2 to 5 microns. Because of their unique strength and extremely low binding characteristics, cellulose acetate membranes are ideal for protein and enzyme filtrations, tissue culture media sterilization, cold sterilization, biological fluid filtration and other filtration applications where maximum recovery of proteins is critical.

Centipoise: A centipoise (cP) is a non-SI (non-System International) measurement unit of dynamic viscosity in the centimeter gram second (CGS) system of units. It is multiple of the CGS base viscosity unit named poise (P). 1 cP = 0.01g/cm/s.

Centrifugation: It is the process of applying centrifugal force to separate fat globules in the form of cream or to separate solid particles from milk and other liquids.

Centrifuge sludge: It is solid and consists of dirt particles, blood particles and foreign matter especially milk protein obtained during clarification. There is a very high load of microorganisms and pathogens in it.

Cerebrosides: Cerebrosides are sphingolipids made by attaching a sugar to a ceramide. Cerebrosides are prominent components of membranes, particularly in the brain. Since they contain a sugar, cerebrosides are also called glycosphingolipids. The most common cerebrosides include galactosylceramide, glucosylceramide, and sulfatides.

Certification: It is a procedure by which a third party gives written assurance that a product, process or service conforms to specific requirements.

Chakka powder: Manufacture of chakka powder involves preparation of dahi followed by partial removal of whey from it. The chakka is then ground in a colloid mill to obtain smooth and uniform consistency in slurry. The slurry containing fat is heated to 38°C and then homogenized (single stage $100 kg/cm^2$) and spray dried with inlet and outlet air temperature of 185°C & 85°C.

Chakka: It is a fermented, intermediate dairy product obtained during the production of Shrikhand. Chakka can be described as strained dahi or curd mass obtained after removing whey from dahi, either through muslin cloth or basket centrifuge. Chakka is the base material for the production of shrikhand and shrikhand wadi.

Cham-cham: It is a traditional Indian dairy product originated in eastern parts of India. It is a sweet prepared from chhana with a firm body and close knit texture. It is filled with sugar or khoa.

Channa: It is India's traditional soft cottage cheese. It is a solid product formed by acid precipitation of milk proteins. It is used as an intermediate base for a wide variety of milk-based Bengali sweets.

Charm analysis: Charm analysis is a bioassay for odor activity in gas chromatographic effluents. It measures the dilution value over the entire time the compounds elute. Hence, it contains the information about the chromatography of odor-active compounds.

Charm test: It is used to determine the antibiotic residue in milk samples. This test is based on specific, irreversible affinity of beta-lactam antibiotics for certain enzyme sites on cell wall of micro-organisms.

Cheddar cheese: Cheddar cheese is a relatively hard, pale-yellow to off-white, sometimes "sharp" (i.e., acidic)-tasting, natural cheese. Originating in the English village of Cheddar in Somerset, South West England.

Cheddaring: It refers to the combined operations of packing, turning, piling and repiling of curd cubes during cheddar cheese making to facilitate removal of whey from the curd.

Cheese analogues: These are the products used as culinary replacements for cheese. These include vegan cheese alternatives as well as some dairy products, such as processed cheese, that do not qualify as true cheeses.

Cheese cultures: These are the selected microbial preparation comprising of desirable group of pure and actively growing micro-organisms, used either singly or in combination, which is intentionally added during cheese manufacture to initiate desirable changes. These microbial preparations can consist of lactic acid bacteria, propionibacteria, surface ripening bacteria and yeasts and moulds.

Cheese curd: The clumps of casein and other milk components that are formed during the cheese making process. These curds are then pressed into blocks or barrels for proper aging and curing of the cheese.

Cheese defects: Lack of something desirable or presence of something undesirable in cheese are known as cheese defects. These defects may be related to color, flavor, body, texture and appearance. Body related defects in cheese include crumbly, corky, mealy, pasty, weak etc. Texture related defects may include openness and gas holes defect in cheese texture. Flavor related defects may include acid, bitter, fruity and unclean. Color related defects may include mottled, seamy and acid cut (bleached). Appearance or finish defects may include greasy, huffed and cracked surface.

Cheese holes (eyes): These are the gas holes formed due to CO2 gas production and are desirable feature in Emmental and other Swiss-type cheeses.

Cheese knives: These are the knives or cutting devices used to cut the coagulum to expel whey and facilitate further operation. These are of two types: horizontal and vertical type. In both types, stainless steel wires are brazed on stainless steel rectangular frame support at a specific distance i.e. 0.5-2.5 cm.

Cheese milk: Milk used for the preparation of cheese is known as cheese milk. It should have a specific casein to fat ratio e.g. for cheddar cheese 0.7 c/f ratio is required.

Cheese milling: It is the process of cutting the cheese curd into shreds or strips using power operated peg mills or by rotating disc cutters before salting and hooping.

Cheese moulds/hoops: These are the perforated containers or equipments used in cheese making to give shape to the curd in hard and semi-hard cheeses during pressing.

Cheese powder: These are the powdered products prepared from cheese using ultrafiltration and drying techniques. Cheese powders can be used for exporting to countries where milk production is very low.

Cheese press: These are the equipments used for pressing of cheese curd to remove whey or moisture from hard and semi-hard varieties of cheeses. These include spring-loaded presses, hydraulic presses, pneumatic presses and vacuum presses. These may be of horizontal or vertical type.

Cheese ripening: Ripening of cheese includes all the physical, chemical, biochemical and microbiological changes occurring in the cheese, during storage of cheese at specific conditions. The structure and composition of the cheese change which leads to changes in its organoleptic properties. Development of cheese properties, including consistency and flavor, is especially attributable to the conversion of lactose, protein, fat, and, in some cheeses, citrate.

Cheese varieties: Cheese can vary greatly in composition and is manufactured and cured in widely varying manners. This results in a bewildering variety of types. Examples include: Cheddar, Cottage, Gouda, Mozzarella, Emmental, Munster, Camembert, Stilton etc.

Cheese vats: These are the double jacketed stainless steel equipments used in processing of milk for cheese making. Generally the vats are rectangular in design with a jacket and special H type arrangement in the bottom for uniform distribution of steam and water supply in the jacket.

Cheese yield: The amount of cheese manufactured from a given weight of milk is known as cheese yield. It is generally expressed as "kg of cheese per 100 kg of milk".

Cheese: It is the product formed by coagulation of milk of certain mammals by rennet or similar enzymes in the presence of lactic acid produced by added or adventitious micro-organisms, from which part of the moisture has been removed by cutting, warming and/or pressing, which has been shaped in mould and then ripened by holding for sometime at suitable temperature and humidity.

Chemical oxygen demand (COD): The amount of oxygen required to oxidize completely the organic matter in water is known as chemical oxygen demand.

Chemical properties: Properties of matter that may only be observed and measured by performing a chemical change or chemical reaction. Chemical properties cannot be determined by touching or viewing a sample; the structure of the sample must be altered for the chemical properties to become apparent. Examples include anti-oxidant capacity, carmelization, acidity, iodine value, saponification value etc.

Chemometrics: It is the science of relating measurements made on a chemical system or process to the state of the system via application of mathematical or statistical methods.

Cherry Burrell method: It is a continuous butter making process in which phase inversion takes place during or immediately after concentration, producing a liquid identical to melted butter, before cooling and working.

Cheshire: It is rennet coagulated, internal bacterial ripened, hard type cheese originated from UK.

Chhana podo: This is baked sweet, resembling milk cake and quite popular in Odhisha. It is prepared by mixing chhana, sugar and semolina or maida. Its surface has caramelized colour and flavour.

Chhana-murki: Chhana-murki is a sugar-coated sweet in the shape of small cubes.

Chhash: Chhash, mattha or chhas are synonyms for a product-buttermilk, which is commonly consumed in all parts of India. It is a popular refreshing summer beverage.

Chocolate milk: These are the milk based products prepared from cocoa or chocolate, sugar and milk with 10% or higher milk fat.

Cholesterol free: Maximum of 2 g saturated fat and less than 2 mg cholesterol are allowed per serving.

Cholesterol: A substance produced in the liver and carried in the blood by two proteins – low and high density lipoprotein. Low density lipoprotein can increase to a high level in the blood; it is deposited in the walls of blood vessels and causes a narrowing of the arteries which can lead to heart attacks and death.

Churned-fat-index: It is a measure of extent of fat clumping during melting of ice cream.

Churning efficiency: Churning efficiency is measured in terms of time required to produce butter granules and by the loss of fat in the buttermilk. Efficiency is influenced by churning temperature and by acidity of milk or cream.

Churning: It refers to the physical agitation of cream until it ruptures the fat globule membrane and then the fat droplets join together to form fat clumps or butter grains. Later, larger clusters of fat collect and form a network with the air bubbles.

Chymosin: It is the enzyme which splits off a portion of the k-casein molecule and degrade its protective ability which leads to precipitation of casein in the presence of Calcium ions.

Clarifier slime: The solid matter, heavier than milk, which is removed by centrifugal clarification. It consists of dirt, body cells, bacteria, hairs, blood cells etc. and is of repulsive odor and appearance.

Clarifier: These are the equipments used for the clarification of products and remove the foreign particles by centrifugal sedimentation. These are similar to centrifugal cream separator in appearance and construction but the discs in the clarifier bowls are smaller in diameter than separator.

Clarifixators: These are the equipments used for inverting phase and separating aqueous serum from liquid milk fat in AMF (or butteroil) manufacture.

Clean: Clean means free of soil, food residue, dirt, grease, cleaning or sanitising agents or other objectionable matter.

Cleaning: It refers to the removal of material that interferes with the proper functioning of processing equipment.

Cleaning-in-place (CIP): Cleaning-in-place refers to the cleaning operations in which the equipment needs not be dismantled and dislocated.

Clotted cream: Cream that is prepared by heating cream to 77-88°C in shallow pans and then allowing it to cool slowly. The surface layer consists of clotted cream which is skimmed off and strained.

Clumping: Clumping refers to the tendency of fat globules to loosely adhere to one another to form clumps.

Clustering: Tendency of small clumps to loosely adheres to one another to form clusters.

CMC: Carboxy methyl cellulose is a food additive which is used for stabilization and thickening of food products.

Coagulants: These are the substances used to coagulate milk specifically the milk proteins. These may include acids (such as lactic acid & citric acid) and enzymes (rennet or chymosin).

Coagulation: A process which involves the denaturation of proteins; changing from a soluble to an insoluble structure. It refers to the liquid milk changing to a semi-solid junket.

Coagulum cutting (cheese): It is the unit operation in cheese making in which curd or coagulum is cut using cheese knives to facilitate whey removal.

Coalescence: It is a type of emulsion instability in which droplets bump into each other and combine to form a larger droplet, so the average droplet size increases over time.

Coarse body (cheese): Cheese defect in which cheese is rough, dry and sandy to the touch.

Coarseness: Coarseness is a quality of roughness or graininess in food products.

Coarsening: It is also known as Ostwald ripening and is an observed phenomenon in solid (or liquid) solutions. When a phase precipitates out of solid, energetic factors will cause large precipitates to grow, drawing materials from the smaller precipitates, which shrinks.

Codex Alimentarius: The Codex Alimentarius Commission was created by FAO and WHO to develop food standards, guidelines and related texts such as codes of practice under the Joint FAO/WHO Food Standards Programme.

Co-extrusion: Co-extrusion is the process of extruding two different materials at the same time in order to form one final unique product.

Coffee cream: It is a type of cream containing 20-25% milk fat. The function of coffee cream is to provide an attractive appearance to the coffee with an appropriate modification in flavor.

Coffee stability: Coffee stability is a measure of a milk powder's resistance to protein instability and resistance to "feathering." Milk protein stability is affected by high temperature, low pH, water hardness (high levels of calcium, magnesium) and other factors.

Coffee whiteners: Coffee whitener or non-dairy creamers are liquid or granular substances intended to substitute for milk or cream as an additive to coffee.

Cold agglutination: Clustering of the milk globules in the cold is referred to as cold agglutination. Creaming in cold raw milk is usually determined by the flocculation of the globules by 'agglutinin,' i.e., a complex of cryoglobulins predominantly immunoglobulin and lipoproteins. In cold agglutination cryoglobulins precipitate onto all kinds of particles, especially fat globules which aggregate into fairly large floccules and rise rapidly.

Cold renneting: An alternative manner of curd making involves concentrating the protein before clotting by means of ultrafiltration. In cold renneting, the concentrate is provided with rennet at a low temperature around 5°C while stirring. Cold renneting greatly shortens the clotting process.

Colloid science: It is the study of interaction forces acting between particles which determine whether particles will aggregate or not.

Colloidal calcium phosphate: Calcium phosphate in milk is found in a colloidal form in micelles bound to casein protein with magnesium, zinc and citrate - collectively referred to as colloidal calcium phosphate (CCP).

Colloidal stability (casein): Casein micelles are lyophilic colloids hence they are physically stable. The micelles do not aggregate under physiological conditions. However, most of the particles are large enough to aggregate owing to van der Waals attraction. Hence, there must be repulsive forces that prevent aggregation. The conditions of milk can be altered in such a way that the micelles do aggregate, for example, by adding considerable quantities of calcium chloride or ethanol or by applying a very high temperature.

Colorants: These are the substances which impart color to the food products. Colorants are characterized by their ability to absorb visible part of electro-magnetic spectrum (380-780 nm).

Colostrum: Colostrum is a milky fluid that is produced by cows, buffalos and other mammals for the first few days after giving birth, before true milk appears. It contains proteins, carbohydrates, fats, vitamins, minerals, and proteins (antibodies) that fight disease-causing agents such as bacteria and viruses. Antibody levels in colostrums can be 100 times higher than levels in regular cow's milk.

Commercial culture: Master culture or the original culture that the dairy buys from the laboratory.

Commercial sterility: A term used in heat sterilisation to indicate that processing inactivates substantially all micro-organisms and spores which, if present, would be capable of growing in the food under defined storage conditions.

Commercial sterility: Commercial sterility means the absence of microorganisms capable of growing in the food at normal non-refrigerated conditions at which the food is likely to be held during manufacture, distribution and storage.

Comminution: Comminution is the reduction of solid materials from one average particle size to a smaller average particle size, by crushing, grinding, and other processes.

Composite milk product: It is a product of which the milk, milk products or milk constituents are an essential part in terms of quantity in the final product, as consumed provided that the constituents not derived from milk are not intended to take the place in part or in whole of any milk constituent.

Compression tests: These are the tests suitable for determining several mechanical and rheological characteristics of different foods with special respects to consumer perception.

Compressive strength: The compressive strength is the capacity of a material or structure to withstand loads tending to reduce size. It can be measured by plotting applied force against deformation in a testing machine.

Computer integrated manufacture (CIM): CIM is the manufacturing approach of using computers to control the entire production process. This integration allows individual processes to exchange information with each other and initiate actions. Through the integration of computers, manufacturing can be faster and less error-prone, although the main advantage is the ability to create automated manufacturing processes

Comte cheese: It is a French cheese made from unpasteurized cow's milk in the Franche-Comte region of eastern France. The rind is usually a dusty-brown colour and the texture is relatively hard and flexible, and the taste is strong and slightly sweet.

Concentrated starter: These are the starters which are concentrated by bactofugation or by microfiltration. These can be freeze dried also.

Condensate: It is the liquid phase produced by the condensation of steam or any other gas.

Condensation: Condensation is the change of the physical state of matter from gas phase into liquid phase, and is the reverse of vaporization. It can also be defined as the change in the state of water vapor to liquid water when in contact with any surface.

Condensed Milk: These are the products obtained by evaporating a part of the water of whole milk or fully or partly skimmed milk, with or without addition of sugar.

Condenser: Condenser is a device or unit used to condense a substance from its gaseous to its liquid state, typically by cooling it.

Confectionery: Confectionery includes the food items that are rich in sugar and carbohydrates. Confectionery is divided into two broad and somewhat overlapping categories, bakers' confections and sugar confections. Bakers' confectionery includes principally sweet pastries, cakes, and similar baked goods. Sugar confectionery includes sweets, candied nuts, chocolates, chewing gum, sweetmeats, pastillage, and other confections that are made primarily of sugar.

Confocal laser scanning microscopy: It is a technique for obtaining high-resolution optical images with depth selectivity. The key feature of confocal microscopy is its ability to acquire in-focus images from selected depths, a process known as optical sectioning.

Consignment: The goods identified in one bill of lading.

Consistency: The property of the material by which it resists permanent change of shape and is defined by the complete force flow relation It is a measure of degree of density, firmness, or viscosity of a milk based product.

Contaminants: A contaminant is any potentially undesirable substance (physical, chemical or biological).

Continuous diafiltration: It involves the continuous addition of water (at the appropriate pH and temperature) in the feed tank at a rate equal to the rate at which permeate is removed, thus keeping feed volume constant during the filtration.

Continuous freezer: These are the freezers used in continuous ice cream manufacturing process and they also contain two pumps to pull ice cream mix to the freezer and from the freezer.

Convergence temperature: It is the temperature at which the contents rest of the freezing–point tube of cryoscope are in thermal equilibrium with the rest of apparatus.

Cooking of curd (cheddar cheese): It refers to the heating of curd cubes. Temperature is raised slowly, if not, a condition similar to 'case hardening' will result and the curd cubes will be hard on outer surface and 'water-logged' inside. The rate of heating is such that the temperature rises to 32°C in about 15 minutes and thereafter to the maximum cooking temperature i.e. 37-39°C at the rate of 1°C every 4 minutes.

Co-precipitate: A co-precipitate of milk proteins may be defined as the product which separates as the solid phase after the heat treatment and precipitation of dairy fluid, or mixtures of dairy fluids, which contain both casein and heat coagulable whey proteins. Co-precipitates are rich in proteins and low in lactose. They can be utilized for the manufacture of foods and beverages for lactose intolerant people or for fortification of infant foods without increasing the lactose and mineral contents.

Corky body (cheese): Cheese defect in which cheese texture that is hard, tough and overly firm.

Corning process or system: It is a system of lactose hydrolysis lactase from *Aspergillus oryzae*, which is covalently bonded to glass beads using a silane-glutaraldehyde coupling.

Corrective action: Action taken to rectify, eliminate the causes of, and prevent recurrence of any problem/ failure/ non-compliance identified in a plan, procedure, process, product, programme, or system.

Cottage cheese: Cottage cheese is a typical lactic acid, unripened, soft curd variety of cheese made from skim milk or from reconstituted concentrated skim milk or non-fat-dry milk solids. A little salt and fresh or cultured cream are generally added to the cheese. It has particulate/granular texture, white or cream yellow color with slight acidic taste and creamy flavor.

Couverture chocolate: Couverture chocolate is a very high quality chocolate that contains extra cocoa butter. The higher percentage of cocoa butter, combined with proper tempering, gives the chocolate more sheen, firmer "snap" when broken, and a creamy mellow flavor.

Cox-Merz rule: The Cox–Merz rule states that the shear-rate dependence of the steady-state viscosity is equivalent to the frequency dependence of the complex viscosity.

Cream aging tanks: The tanks are usually vertical with cooling panels or coils built into the walls. Agitation is by slow-moving gate type agitators which may also contain refrigerant.

Cream concentrators: These are the equipments used to produce plastic cream (60-80% fat) from normal cream of approximately 40% fat.

Cream layer or line: The fat, being lighter than milk, slowly accumulates at the surface to form the cream layer, commonly called the cream line.

Cream line index: It is the ratio of percentage cream line (layer) to the percentage of fat in the milk.

Cream liqueurs: These are the alcoholic beverages and represent a substantial volume market for cream in western countries. The alcohol acts as a preservative against bacterial spoilage and the resultant products have long shelf-lives if suitable steps are taken to prevent creaming and destabilization of the emulsion. Hence, homogenization along with sodium caseinate and sodium citrate addition is important to provide emulsion stability.

Cream neutralization: Neutralization of sour cream for butter making refers to a partial reduction in its acidity. Two types of neutralizers used are lime and soda. It avoids excessive fat loss in buttermilk, guard against production of undesirable off-flavors in cream and improve the keeping quality of butter made from high-acid cream.

Cream powder: Cream powder or dried cream refers to a dried milk product which has a higher fat content than whole milk powder.

Cream separation (centrifugal): Centrifugal cream separation is used to make cream and skim milk, to obtain some cream from whey or sweet-cream buttermilk, and to standardize milk and milk products to a desired fat content.

Cream separator: A cream separator is a centrifugal device that separates milk into cream and skimmed milk. It is virtually a series of conical plates which revolve in bowl at high speed and fling the debris of milk outwards and the fat being the lightest fraction collects in the centre from which it is collected and led out.

Cream: Cream is a dairy product that is composed of the higher-milk fat layer skimmed from the top of milk before homogenization. In un-homogenized milk, the fat, which is less dense, will eventually rise to the top.

Creaming: It is the process of rise of fat globules in milk due to the difference in density between milk plasma and fat globules. It enables the milk to be separated into cream and skim milk.

Creep compliance: The ratio of strain at time *t* to the constant load is called 'creep compliance'.

Creep: The increase in deformation or strain as a function of time is known as creep.

Critical Control Point (CCP): It is a point, step or procedure at which controls can be applied and a food safety hazard can be prevented, eliminated or reduced to acceptable (critical) levels.

Critical moisture content: The amount of moisture in a food at the end of the constant-rate period of drying.

Critical point: A critical point is defined as one that must be controlled to ensure safety or good quality. There may be several other points (process steps or product properties) that can be, but need not be, controlled because control of another, critical, point will also detect the imperfection.

Crumbliness: The word derived from the term crumb and therefore it refers to the quality of being excessively breakable. It also refers to the regularity of something that breaks under pressure such that it is easily crumbled; friable.

Crumbly body (cheese): Cheese defect in which cheese is loosely knit and tends to fall apart when rubbed between the thumb and fingers.

Cryogen: A refrigerant that absorbs latent heat and changes phase from solid or liquid to a gas, e.g. subliming or evaporating carbon dioxide or liquid nitrogen.

Cryogenic freezers: Equipment that uses subliming or evaporating carbon dioxide or liquid nitrogen directly in contact with food to freeze it.

Cryogenic freezing: Cryogenic freezing is a type of freezing which requires extremely low temperatures, generally below-150°C. This process is part of a branch of the sciences known as cryogenics, which focuses on the production of very cold temperatures and the study of what happens to objects subjected to these temperatures.

Crystallization: It is the process of formation of solid crystals precipitating from a solution, melt or more rarely deposited directly from a gas. Crystallization is also a chemical solid–liquid separation technique, in which mass transfer of a solute from the liquid solution to a pure solid crystalline phase occurs.

Cultured cream: It is a type of cream which is soured and thickened by adding lactic acid bacteria to pasteurized cream.

Curd fines: Loss of curd particles from cheese curd into whey during curd treatment such as cutting and stirring. Loss of curd fines greatly depend on the firmness of the renneted milk at the moment of cutting and on the intensity of cutting and stirring.

Curd fusion (cheese): During molding, the curd grains are sufficiently deformable to enable them to fuse into a coherent mass. This property is known as curd fusion.

Curd knives: These consist of stainless steel wires or strips 6 or 9 mm apart, one is horizontal and other is vertical.

Curd pressing (cheese): This refers to the operation of forcing the particles of milled and salted curd in the hoops into the smallest possible space.

Curdling: It refers to liquid milk changing to a semi-solid junket.

Curdy body (cheese): Cheese defect in which cheese is smooth but firm, and when worked between the fingers, is rubbery and not waxy.

Cutting (cheese): It refers to cutting of the firm coagulum into cubes of a specific size.

Cyclone collector: These are the equipments used to remove heavy particles from the airflow.

D

Dacca cheese: It is the cheese prepared from buffalo milk treated with rennet and salt.

Dahi: Dahi or curd is the product obtained from pasteurized milk by souring by a harmless lactic acid or other bacterial culture. It may contain additional cane sugar.

Dahiwada: It is eaten as a snack or may accompany a meal as a side dish. To prepare dahiwada, deep fried black gram *dal* batter patties (*wada*) are dipped in *dahi* and allowed to soak. They are usually garnished with sweet *chutney* prepared from tamarind and jaggery.

Dairy plant: The term is generally used for all engineering equipment, especially that for pasteurizing, cooling, bottle washing, filling and capping, steam raising, refrigeration etc.

Dairy spreads: Dairy spreads are milk products relatively rich in fat in the form of a spreadable emulsion principally of the type of water-in-milk fat that remains in solid phase at a temperature of 20°C.

Dariloid: It is the stabilizer (sodium alginate) used in ice cream sold under the trade name of 'Dariloid'. It is extracted from seaweed. It improves the whipping ability and leaves a slightly cleaner flavor in the mouth. It dissolves properly only when added to the mix at about 68-71°C.

Decimal reduction time: It refers to D-value and is the time required at a certain temperature to kill 90% of the organisms being studied.

Decreaming: It is the mechanical separation of milk into cream and skim milk by means of centrifugal forces.

Dehydration: It is one of the oldest methods of food preservation by removal of moisture from the food. It increases the shelf life of food by inhibiting the growth of microorganisms.

Demineralization: It is the process of removal minerals from a liquid (such as whey) by ion exchange, reverse osmosis and electrodialysis.

Denaturation of Proteins: Loss of tertiary structure of proteins or unfolding of peptide chains under the effect of heat and acid.

Densitometer: These are the equipments used for determining fat content of cream using density.

Density: Mass per unit volume of substance is known as density.

Descaling: The removal of milk stone (calcium salts) or of rust and other corrosion oxides etc., from metals by acids.

Descriptive analysis: It is the discipline of quantitatively describing the main features of a collection of information. It aims to summarize a sample, rather than use the data to learn about the population that the sample of data is thought to represent.

Detergents: Chemicals that reduce the surface tension of water and hence assist in the release of soils from equipment or foods.

Deterioration: It may include losses in organoleptic desirability, nutritional value, safety and aesthetic appeal. Food may subject to physical, chemical and biological deterioration.

Dew point: Temperature at which an air–water vapours mixture becomes saturated with moisture, marking the onset of condensation.

Dextrose equivalent (DE): It is a measure of the reducing sugar content of the syrup calculated as dextrose and expressed as percentage of the total dry weight.

Diafiltration: It is the process of adding water to the retentate so that further removal of membrane-permeable solutes can continue along with the added water. It is of two types: discontinuous diafiltration and continuous diafiltration.

Die: A restricted opening at the discharge end of an extruder barrel.

Dielectric constant: The ratio of the capacitance of a food to the capacitance of air or vacuum under the same conditions.

Differential scanning calorimetry (DSC): It is a technique for measuring the thermal behavior of a substance. In DSC, the difference between the heat flow (J/s, or W) to or from a sample and a reference material is measured as a function of temperature or time while the sample and the reference material are subjected together to a controlled temperature-time programme.

Digestion: It is a form of catabolism which involves mechanical and chemical breakdown of food into smaller components that are more easily absorbed into a blood stream.

Dipping (cheese): It refers to the removal of whey from the curd.

Direct digital control (DDC): The measurement signals are connected to analogue inputs, the modulating valve is connected to an analogue output, and the set-points and tuning constants are entered through the operator's workstation. Implement of direct control of output is necessary when a process has complex start-up and shutdown requirements. It is often termed as direct digital control (DDC).

Direct vat inoculation: It refers to the direct inoculation of milk with an adequate number of lactic acid bacteria by means of a deep-frozen or freeze-dried concentrated bulk starter culture with up to 10^{13} CFU per gram dispenses with the preparation of a bulk starter.

Discontinuous diafiltration: It refers to the operation where permeable solutes are cleared from the retentate by volume reduction, followed by re-dilution with water and repeated volume reduction in repetitive steps.

Disease: Any illness which has the potential to make the raw milk unfit for the manufacture of dairy products for consumption.

Disinfection: It is the process of killing pathogenic organisms or rendering them inert.

Disintegration (micelles): Weakening of the bonds between the submicelles or those between protein molecules in the submicelles can lead to disintegration of the micelles.

Dispersibility: It refers to the dispersion of milk powder in a defined time at 25°C (room temperature). It primarily depends on the rate at which water fully penetrates a mass of powder. If it happens quite fast, the powder is said to have 'instant' properties.

Domiati: It is rennet coagulated, internal bacterial ripened, high salt cheese variety originated from Egypt.

Dornic degrees: One Dornic degree corresponds to 0.01g lactic acid per 100ml of milk or 0.01 percent. Therefore, the number of Dornic degrees divided by 100 gives the content of lactic acid in percent.

Dornic method: It is a method used for determining acidity of milk and milk products.

Double cream: A thick cream of high fat content e.g. 50%.

Double toned milk: It refers to milk obtained by the addition of water and skim milk powder to the whole milk. It should contain a minimum of 1.5% fat and 9.0% SNF.

Dressing (cheese): This refers to the arrangement of the cheese cloth before and after pressing.

Drum drying: It is a method used for drying out liquids. In drum drying, milk is applied as a thin film to the surface of a heated drum, and the dried milk solids are then scraped off with a knife. Powdered milk made by drum drying tends to have a cooked flavor, due to caramelization caused by greater heat exposure.

Dry salting (cheese): In this process, salt crystals are mixed with the curd grains or with the milled curd pieces resulting from traditional cheddaring.

E

Early blowing (cheese): Cheese defect usually occurs within 48 hours of the cheese making and appears as small, spherical, shiny holes. The defect is often associated with unclean flavour.

Economiser: It is a component of a spray drier designed to cool the flue gases from the oil burner from about 500°C to 180°C by means of return fluid. (160°C) from the air heater.

Edam cheese: It is named after the Edam town in Holland. It is a semi-hard to hard rennet curd, bacterial ripened cheese having few small (2-10 mm size), round to oval gas holes (eyes), having very mild, not acid, clean flavor. Its texture is elastic, pliant, and softer than Gouda. Its shape is mostly flattened spherical but may also be square blocks. Its surface is coated with wax.

Edible casein: Edible casein is defined as the casein which has been isolated from skim milk by taking special precautions to ensure its suitability for use in patented food and pharmaceutical preparations. These precautions include: strict control of quality of raw material; use of standard equipment and technique for production; packaging and storage under approved conditions.

Effect: A single unit in an evaporator operating at a particular pressure and temperature. Evaporators commonly have three to seven effects to allow heat to be reused several times.

Effluent: The liquid discharge from sewage treatment and industrial plants.

Egg yolk: It refers to the portion of egg which is used as natural emulsifier in ice cream and other dairy products. Fresh egg yolk contains about 33% fat and 50% solids.

Elastic behaviour: The material is said to be have elastic behavior when the material will return to its original shape on release of stress.

Elastic material: These are the materials that regains it shape and geometry after removal of external forces. In a purely elastic material, all energy added is stored in the material.

Elastoplastic behaviour: When a substance has both the elastic as well as plastic properties then it is known as elastoplastic behavior.

Electrodialysis: It is a membrane process based on electrochemical laws governing the migration of ions under the effect of direct current at a given potential through an ion-selective semi permeable membrane.

Electrophoresis: It is defined as the migration of charged molecules under the influence of an electric field. Gel used in electrophoresis is generally polyacrylamide gel.

Emmental cheese: Emmental is a ripened hard cheese, usually manufactured as wheels and blocks of weights from 40 kg or more, adapted to various use conditions. The body of the cheese has an elastic texture, with regular cherry to walnut sized gas holes, or eyes. The activity of propionic acid bacteria is essential to eye formation.

EMP pathways: It is a metabolic pathway which converts glucose into pyruvate.

Emulsification: Creation of an emulsion by the dispersion of one immiscible liquid (dispersed phase) in the form of small droplets in a second immiscible liquid (continuous phase).

Emulsifiers: These are the substances which stabilize the emulsion and consist of water loving hydrophilic head and oil loving hydrophobic tail. The hydrophilic head is directed to the aqueous phase and the hydrophobic tail to the oil phase

Emulsion Stability: It refers to the ability of an emulsion to resist any change in its properties over time. There are four types of instability in emulsions: flocculation, creaming, coalescence and Ostwald ripening.

Emulsion: An emulsion is a mixture of two or more liquids that are normally immiscible. These are of two types: oil-in-water and water-in-oil.

Endopeptidases: These are the proteolytic peptidases that break peptide bonds of non-terminal amino acids i.e. within the molecule e.g. trypsin, pepsin etc.

Enrobing: The unit operation in which food pieces are coated with chocolate or other materials.

Enter-Doudoroff pathway: Glycolytic pathway that results in the net production of only one ATP molecule per molecule of glucose substrate metabolized.

E-number: An E number is a reference number given to food additives that have passed safety tests and have been approved for use throughout the European Union (EU), for example, pectin (E440), sorbitol (E420).

Enzyme: These are the complex proteins which help in the biological reactions and hence also known as biocatalysts.

Enzyme-modified cheese: Enzyme-modified cheese (EMC) is cheese curd which has been treated with enzymes to produce a concentrated cheese flavour ingredient.

Epilactose: It is a disaccharide composed of galactose and mannose.

Ergosterol: A sterol which is activated by sunlight to become vitamin D.

EU: European Union is an intergovernmental and supra-national union of 25 democratic member states from the European continent.

Eutectic temperature: The temperature at which a crystal of an individual solute exists in equilibrium with the unfrozen liquor and ice.

Evaporated milk: The product obtained by partial removal of water from milk of cow and/ or buffalo by heat or any other process which leads to a product of the same composition and characteristics. It should not contain less than 8.0% milk fat and not less than 26% milk solids.

Evaporation: The changing of liquid into gas. Considerable heat or energy is required for this phenomenon, to free the molecules from the liquid phase. This energy is known as latent heat of evaporation.

Exoenzymes: Enzymes that occur either attached to the outer surface of the cell membrane or in the periplasmic space.

Explosion doors: It consists of a hinged cover with a closing device and a seal between the cover and the vessel to which it is mounted. The door is normally equipped with a buffer device to catch the cover when it is thrown out during the explosion.

Explosion pressure relief: An explosion pressure relief device allows an unburned dust-air mixture and the combustion gases to be led out from the components where the explosion takes place.

Explosion suppression: An explosion suppression system primarily consists of explosion detectors, a number of pressure bottles containing a suppression agent and a central control unit. It registers the explosion in its initial phase then immediately injects the suppression agent to reduce the explosion pressure.

Export: Transport the goods outside the country.

Eyes (cheese): Holes produced by the propionic acid fermentation.

F

Falling film evaporator: This is a type of evaporator in which the liquid flows by gravity along the inside surfaces of a number of pipes as a thin film; the heating medium is at the outside. The liquid passes a pipe only once and drops from the pipe in a concentrated form. A difficulty is that the pipes should be evenly and completely wetted by the liquid to avoid loss of heating surface and excessive fouling.

FAO: Food and Agriculture Organisation. FAO is a specialized agency of the United Nations that works to raise levels of nutrition and standards of living; to improve the production, processing, marketing, and distribution of food and agricultural products; to promote rural development; and, by these means, to eliminate hunger.

Fat bloom: It is similar to sugar bloom, except that it is fat or cocoa butter that is separating from the chocolate and depositing itself on the outside of the candy.

Fat crystallization: It is a complicated phenomenon, especially for milk fat with its very broad composition. Fat crystals greatly affect consistency and mouth feel of high-fat products, especially butter; and the physical stability, especially to partial coalescence, of milk fat globules.

Fat crystals: When fats are cooled down from a high temperature (so completely melted fat) to temperatures below the melting point, crystals are formed. Different crystal modifications can be formed such as alpha-solids which are formed when fat is quickly cooled down in seconds; these solids are unstable and will convert in more stable form. Beta-prime solids and Beta solids; these are the most stable modification.

Fat destabilization: Fat destabilization refers to the process of clustering and clumping (partial coalescence) of the fat globules which leads to the development of a continuous internal fat network or matrix structure in the product.

Fat globule membrane: The native fat globule membrane (FGM) is comprised of apical plasma membrane of the secretory cell which continually envelopes the lipid droplets as they pass into the lumen. The major components of the native FGM, therefore, are protein and phospholipids.

Fat globules: More than 95% of the total milk lipid is in the form of a globule ranging in size from 0.1 to 15 um in diameter. These liquid fat droplets are covered by a thin membrane, 8 to 10 nm in thickness, whose properties are completely different from both milk fat and plasma. The native fat globule membrane (FGM) is comprised of apical plasma membrane of the secretory cell which continually envelopes the lipid droplets as they pass into the lumen. The major components of the native FGM, therefore, are protein and phospholipids.

Fat in dry matter (FDM): It represents the fat content in the dry matter of the product. The FDM content is between 42–56% for most of the rennet-curd cheese varieties e.g. Cheddar, Gouda, and Blue.

Fat oxidation: Fat oxidation is the cause of important deteriorative changes in chemical, sensory and nutritional food properties.

Fat replacers: These are the non-fat sources which provide fat-like properties when added to different food and dairy products e.g. gums, mucilages etc.

Fat: Fats consist of a wide group of compounds that are generally soluble in organic solvents and generally insoluble in water. Chemically, fats are triglycerides: triesters of glycerol and any of several fatty acids. Fats may be either solid or liquid at room temperature, depending on their structure and composition.

Fatty acid: A carboxylic acid with an unbranched hydrocarbon chain having about 14-24 carbon atoms.

Feathering of cream: A term applied to flocculation of fat etc., on the surface of tea or coffee, particularly with homogenized milk. The most important factors appear to be the acidity and pressure of homogenization.

Feed pumps: These are the sanitary, volumetric pump with a variable speed drive.

Fermentation: It is a process of conversion of carbohydrate into acid or alcohol by the action of one or more micro-organisms in specific conditions of temperature, pH, water activity and nutrients.

Fermented milks: Fermented milks refer to those milks which have been made by employing selected micro-organisms to develop the characteristic flavor and/or body and texture e.g. Kefir, Kumiss etc.

Feta cheese: Feta-type cheeses are first salted, after which they are packaged and cured in brine or in acid whey with salt added.

Filled milk: It refers to the product obtained when vegetable fat, skim milk powder and water are combined in correct proportions to yield fluid milk.

Filtermat drier: A Filtermat drier has a continuous woven belt for collecting powder from the primary chamber and cool air can then be directed on to the powder in subsequent (fluid bed) stages of drier. This drier is extensively used to dry powders with high fat content.

Final eutectic temperature: The lowest eutectic temperature of solutes in equilibrium with unfrozen liquor and ice.

Finished product: Dairy material or product which has been packed, in the manner intended for sale, and is awaiting the decision concerning conformance with regulatory requirements.

Firm body (cheese) - Cheese defect in which the body of the cheese feels solid and is not soft or weak.

Flash pasteurization: Pasteurization requiring very short times is known as 'flash pasteurization'. For example at 77°C the extrapolated time would be about one second.

Flat flavor (cheese defect): An insipid flavour or one which is practically devoid of any characteristic cheese flavour for the applicable variety.

Flat sour: A canned product (e.g. evaporated milk) which sours without the production of gas.

Flavor enhancer: Substance added to food in order to enhance or intensify its flavor. Salt is a common flavor enhancer.

Flavor perception: It is a highly complex organoleptic phenomenon. Traditionally, two chemical senses are distinguished: taste and odor, which are sensitive to specific water-soluble and volatile compounds, respectively. It has long been known that a combination of taste and odor leads to perceived flavor. Flavor perception is significantly dependent on one's experience with various foods and on the conditions during flavor assessment.

Flavor suppressor: These are the substances which suppresses the flavors such as bitterness in food products. Examples of flavour/bitterness suppressor include: sucrose, citric acid.

Flavoured milk: It is a type of fluid packaged milk in which flavouring has been added, such as chocolate, strawberry and vanilla.

Flavourings: These are any of the liquid extracts, essences, and flavours that are added to foods to enhance their taste and aroma. Flavourings are prepared from essential oils, such as almond and lemon; from vanilla; from fresh fruits by expression; from ginger by extraction; from mixtures of essential oils and synthetic organic chemicals; or entirely from synthetic chemicals, with alcohol, glycerol, propylene glycol, alone or in combination, as solvents. Water is added and sometimes certified food colour as well.

Flocculation: It is a type of emulsion instability. It occurs when there is an attractive force between the droplets hence they form flocs, like bunches of berries.

Flow diversion device: It is also known as the flow diversion valve (FDV), it is located at the downstream end of the upward sloping holding tube. It is essentially a 3-way valve, which, at temperatures greater than 72°C, opens to forward flow. This step requires power. At temperatures less than 72°C, the valve recloses to the normal position and diverts the milk back to the balance tank. It is important to note that the FDD operates on the measured temperature, not time, at the end of the holding period. There are two types of FDD: single stem and dual stem.

Flow wrapping: It is a type of packaging in which wrapping material is formed into a tube and the finished product is inserted into it. The tube is then cut and heat sealed. Flow wrapping is faster (than traditional packaging), reduces packaging material cost and allows the continuous production of individually wrapped product.

Flowability: The capacity to move by flow that characterizes fluids and loose particulate solids is known as flowability. This characteristic is a measure of the free-flow characteristic of a powder. Proper flow of milk powders is important for the manufacturer and end-users for proper packaging, handling and measuring. In general, powders with good flow properties are those with large agglomerates and few fines.

Fluid bed: Equipment used for drying or cooling milk powder. Air is blown through the powder from below, causing the powder particles to separate and behave rather like a fluid. Alternatively, a layer of fluid-like powder in which the particles are kept apart by an air flow.

Fluidity: The property of flowing possessed by gases and liquids.

Fluorescence: The emission of radiation in the visible range of a different wave-length from that falling on a substance. The incident radiation may or may not be visible.

Foam drying: It is a form of drying in which air or nitrogen is injected into the concentrate under pressure and the mixture obtained is heated in a vacuum.

Foaming: Skim milk foams readily generally at low temperatures. The foam is subject to drainage, Ostwald ripening, and coalescence, and has a lifetime of about an hour. But at temperatures around 40°C, the lifetime is longer (upto 5 h). Milk fat globules depress foaming.

Fobbing: This is the term used in the trade to describe foaming, especially in relation to foaming of detergent solutions in bottle-washing machines.

Folic acid: It is a form of water-soluble vitamin B_9. Folate is a naturally occurring form of the vitamin, found in food, while folic acid is synthetically produced, and used in fortified foods and supplements.

Food infection: When foods containing pathogenic bacteria are consumed and the bacteria are able to colonize in the intestinal tract of the host, grow and cause tissue damage, the resultant symptoms are referred to as food infection.

Food intoxication: It is also known as true food poisoning and is the direct result of ingesting food containing a toxin produced during bacterial growth in thc food.

Food poisoning: Food poisoning is an illness caused by eating contaminated food, pathogenic bacteria, viruses or parasites as well as chemical or natural toxins.

Fore-milk: The first milk to be withdrawn from the udder at the beginning of milking.

Form fill sealing machine: Form fill sealing machine (FFS) is a type of automated assembly-line product packaging system, commonly used in the packaging of milk, butter-milk, ghee etc. The machine constructs plastic bags out of a flat roll of plastic film, while simultaneously filling the bags with product and sealing the filled bags.

Formalin: It is a 40% solution of formaldehyde and used as a preservative for milk.

Fouling: Deposits of food or lime scale on surfaces of heat exchangers.

Free fatty acids: FFAs are used to estimate the immediate quality of oils and fats and products containing oil or fat. Good quality oil has a low content of free fatty acids.

Free-flowingness: It refers to the ability of a powder to be poured. This property may be estimated by pouring out a heap of powder under standardized conditions and measuring its angle of repose.

Freeze drying of milk: It consists of freezing the product and supplying heat, so that moisture is removed by sublimation by maintaining a vacuum in the vaporizing chamber. Freeze-dried milk consists of coarse, irregularly shaped, and very voluminous powder particles, which dissolve readily and completely.

Freeze drying: It is a drying process which involves the freezing of material and then reducing the surrounding pressure to allow the frozen water in the material to sublimate directly from the solid phase to the gas phase. It is also known as lyophilisation.

Freezing: Freezing is a common method of food preservation that slows both food decay and the growth of micro-organisms. It may also refer to phase transition in which a liquid turns into a solid when its temperature is lowered below its freezing point.

Frequency: The number of complete oscillations or vibrations per second in a current or vibrating material or medium.

Fritz method: This is a continuous method of butter production. It involves the use of high-speed beaters to destabilize the fat emulsion in the chilled cream, and thus cause the formation of butter grains in a matter of seconds. The butter milk is drained away and the resulting grains worked in a kneading section prior to extrusion.

Fromage Frais: Fromage Frais is a soft, unaged, creamy fresh cheese made out of whole or skimmed milk and cream. A French specialty, traditionally this cheese is made with unpasteurized cream.

Frozen yoghurt: It is a frozen dessert made with yogurt and sometimes other dairy products. It varies from slightly too much more tart than ice cream, as well as being lower in fat. It may or may not contain live and active bacteria culture. Frozen yogurt usually consists of milk solids, some kind of sweetener, milk fat, yogurt culture (*Lactobacillus bulgaricus* and *Streptococcus thermophilus* are common cultures), natural or artificial flavorings, and sometimes natural or artificial coloring.

Fructose: It is a simple ketonic monosaccharide found in many plants and fruits and is also known as fruit sugar. Pure, dry fructose is a very sweet, white, odorless, crystalline solid and is the most water-soluble of all the sugars. From plant sources, fructose is found in honey, tree and vine fruits, flowers, berries, and most root vegetables.

FSSAI: The Food Safety and Standards Authority of India (FSSAI) is an agency of the Ministry of Health & Family Welfare, Government of India. The FSSAI is responsible for protecting and promoting public health through the regulation and supervision of food safety. The FSSAI has been established under the Food Safety and Standards Act, 2006 which is a consolidating statute related to food safety and regulation in India.

Functional food: A functional food is a food given an additional function (often one related to health-promotion or disease prevention) by adding new ingredients or more of existing ingredients. These active ingredients may include vitamins, dietary fibers, probiotics etc.

Functional properties: These are the food properties that determine the end-use of a food product. It characterizes the structure, quality, nutritional value and acceptability of a food product. A food functional property is determined by physical, chemical, and/or organoleptic properties of a food. Examples of a functional property may include solubility, absorption, water retention, frothing ability, elasticity, absorptive capacity etc.

***F*-value:** The time required to destroy a given percentage of micro-organisms at a specified reference temperature and *z* value.

G

Galactan: Galactan is a polymer of the sugar galactose found in hemicellulose. It is a gum-like carbohydrate material formed by certain bacteria and is a cause of ropiness in milk.

Galactose: It is a monosaccharide sugar that is less sweet than glucose and fructose. It is a hexose forming half of the molecule of lactose.

Gangliosides: The gangliosides are complex ceramide polyhexosides, which contain one or more acidic sugars [known as sialic acid or N-acetylneuramic acid (NANA)]. The specific names of the gangliosides identify their structure.

Gasohol: A mixture of gasoline and ethanol used as a fuel.

Gassy texture (cheese) - Cheese defect in which cheese contains gas holes of various sizes which may be scattered or unevenly distributed throughout the cheese.

Gel: It refers to dilute cross-linked system, which exhibits no flow when in the steady state. Gel is a solid, jelly-like material that can have properties ranging from soft and weak to hard and tough. The coagulum in cheese making is an excellent example.

Gelatine: It is formed from collagen upon heating and is used as an additive in various dairy products.

Gelation: The process of formation of space filling network or gel is known as gelation. Freeze gelation refers to the solidification by cooling or freezing.

Ghee making: It is an important cottage industry in India. In ghee making, butter obtained from curd or fermented cream is heated in a pan over a low fire. At about 90°C, it begins to boil when it is gently stirred. After most of the water has boiled away, the temperature rises and at about 120°C the curd particles begin to brown and a pleasant aroma of caramelization emanates. This is taken as the end of the clarification process. The fat layer is then carefully removed from solid curd or by straining ghee through a muslin cloth or a strainer and allowed to cool and crystallize. In the direct cream process, cream obtained from milk is mixed with an equal volume of water and reseparated. It is then heated as in the case of butter to obtain ghee.

Ghee residue: It is a by-product of ghee industry. It refers to the charred (burnt) light to dark brown residue which is obtained on the cloth strainer during filtration of ghee after heat clarification.

Ghee: Ghee may be defined as clarified butter fat prepared chiefly from cow or buffalo milk

Glass transition: It takes place without the release or absorption of latent heat and involves the transition of a food to an amorphous glass state at its glass transition temperature. The transition is dependent on the temperature of the food, time, and the moisture content of the food.

Glycemic index: Glycemic Index (GI) is a measurement carried out on carbohydrate-containing foods and their impact on our blood sugar. It is a measure of how quickly blood glucose levels rise after eating a particular type of food. Glucose (the defining standard) has a glycemic index of 100.

Glycerol monostearate: Commonly known as GMS and is an organic molecule used as an emulsifier. GMS is a colorless, odorless, and sweet-tasting flaky powder that is hygroscopic. It is a glycerol ester of stearic acid. GMS is a food additive used as a thickening, emulsifying, anti-caking and emulsifying agent.

Gorgonzola cheese: It is rennet coagulated, internal mould ripened (*P. roqueforti)* cheese variety originated in Italy.

Gouda cheese: It is a hard-pressed variety of cheese of Dutch origin. It is a sweet-curd renneted cheese made from partly defatted milk. The curd is cooked with little acid development. The presence of some shiny 'eyes' is normally observed in this cheese.

Gover's process: A type of vacuum roller drier with two drums operating on the ordinary principle.

Grading of milk: It refers to the classification of milk on the basis of quality, for price-fixing purposes. The principle of grading is based on the platform tests.

Grafset: It is a design tool to overcome many of the difficulties of configuring ladder logic applications. The application programme is designed graphically by specifying a number of operations or states and defining the logic which causes the transition from one state to another.

Grana cheese: It belongs to the family of extra hard cheese which is rennet coagulated and internal bacterial ripened. It has a granular structure and fine sweet, mellow, nut like flavor that is contributed by free fatty acids and amino acids.

GRAS: Generally regarded as safe, typically used in the US. Can be self GRAS or given by FDA

Grating (Cheese): It is the process of grating/ shredding the cheese (hard) into fine pieces using grater.

Gravity separation: It is an industrial method of separating two components, either a suspension, or dry granular mixture where separating the components with gravity is sufficiently practical: i.e. the components of the mixture have different specific weight. The instruments used for gravity separation are known as gravity separators.

Gruyere cheese: It is rennet coagulated, internal bacterial ripened, Swiss-type cheese having eyes which are caused by lactate metabolism by *Propionibacterium* spp.

Guar gum: It is a galactomannan obtained from the seed endosperm of guar plant and is composed of galactose and mannose. It is used as stabilizer and thickener in various dairy and food products.

Gulab jal (rose essence): It is a most favoured form of rose essence among Indian sweets makers. It is a clear liquid, distilled from fresh rose petals. It is a flavouring that can be subtle, when used lightly or soapy when added in larger quantities.

Gulabjamun: Gulabjamun is a khoa based sweet popular all over India. According to BIS specifications, it shall be prepared from khoa, skim milk powder, milk powder, ghee, cream, butter or other milk solids. It may contain maida (wheat flour), citric acid and baking powder. It shall be free from dirt and other foreign matter as well as insects and mould growth. It should be, as far as possible, manufactured and packed under hygienic conditions. The proportion of free syrup in a gulabjamun pack shall not exceed 60% of the declared net mass.

Gums: These are viscous, adhesive, complex polysaccharides or carbohydrates. Certain bacteria can produce gums in milk and so cause ropiness.

H

HACCP: HACCP is the systematic assurance of the hygienic quality of foods in order to exclude risk to the health of consumers. It may also be defined as systematic identification of hazards and the measures for their control to ensure the safety of food. It focuses on prevention rather than end-product testing.

Handling of milk: The term is used to describe the various operations by which milk is collected, pasteurized, stored and distributed.

Hard water: It is the water that contains salts of calcium and magnesium principally as bicarbonates, chlorides, and sulfates. Water hardness that is caused by calcium bicarbonate is known as temporary, because boiling converts the bicarbonate to the insoluble carbonate; hardness from the other salts is called permanent.

Hazard analysis: The identification of potentially hazardous ingredients, storage conditions, packaging, critical process points and relevant human factors which may affect product safety or quality.

Head space/Air space: It refers to the empty space left above the contents in a sealed container or unfilled space in a container, particularly containing a liquid.

Heat acid coagulation: This process permits recovery of caseins and whey proteins in a single step. The basic principle is that whey proteins which are normally acid stable become sensitive to acid coagulation after heat treatment.

Heat stability (milk): Heat stability of milk may be defined in terms of the time required to induce coagulation at a given temperature, such as 115°C. Heat stability of evaporated milk refers to the relative resistance of the milk to coagulation in the sterilizer. Caseins and salt balance of milk governs its heat stability. Added citrates, phosphates and calcium have a great impact on the heat stability.

Hedonic scaling: It is also called acceptability scaling. It measures the degree to which people like the food product. The most common hedonic scale is a 9-point hedonic scale.

Henry's law: It is a gas law which states that, at a constant temperature, the amount of a given gas that dissolves in a given type and volume of liquid is directly proportional to the partial pressure of that gas in equilibrium with that liquid.

Hermetic separator: These are the separators which are based on the provision of hermetic seals which effectively isolate the separator from the atmosphere as entrainment of air in milk inhibits separation.

High-heat SMP: SMP which requires a more intense heat treatment in addition to pasteurization. This product is used in the bakery industry, chocolate industry, and other foods where a high degree of protein denaturation is required.

Holding time test: The holding time of a HTST pasteurizer is the flow time of the fastest particle of milk at a prescribed temperature through the holding section. The holding time is calculated between the points at which the heated milk leaves the heating section and reaches the FDV.

Homogalactouronan: It is a polysaccharide made up of linked galacturonic acid.

Homogenization: It is the treatment given to milk which causes disruption of milk fat globules into smaller ones.

It also enlarges the milk fat–plasma interface usually by a factor of 5 to 10.

Homogenized milk: Milk that has been treated to ensure breakup of fat globules. Homogenization prevents the cream portion of milk from separating from the skim portion. A test of adequate homogenization is such that after 48 hours of quiescent storage at room temperature, no visible cream separation occurs on the milk.

Homogenizer: Homogenizers are generally consists of a high-pressure pump that forces the liquid through a narrow opening called homogenizer valve.

HTST: High-temperature, short–time pasteurization (~72°C for 15 seconds). It is the most commonly used process for pasteurizing milk.

Huffed texture (cheese) - Cheese defect in which cheese is swollen because of gas fermentation and has become rounded oval in shape rather than flat.

Hydraulic retention time: It is a measure of the average length of time that a soluble compound remains in a constructed bioreactor. Hydraulic retention time is the volume of the aeration tank divided by the influent flow-rate. It is also known as hydraulic residence time.

Hydrocolloid: These are the most widely used additives; they are macromolecular and hydrophilic in nature. They are soluble in water, can be dispersed and absorb water and result in highly viscous solutions at very low concentrations. They are used in emulsions and foams to modify appearance, texture, stability and taste due to their ability to form gels or to increase the viscosity of aqueous solutions.

Hydrogenation: Hydrogen is added to carbon - carbon double bonds in unsaturated products in the presence of a catalyst. This results in a harder product (increased melting point). All double bonds can be converted to saturated bonds (fully hydrogenated) or part of the double bonds can be converted to saturated bonds (partially hydrogenated).

Hydrolysis: Hydrolysis is a reaction in which a molecule is split into two parts by reacting with a molecule of water. For example fat will be hydrolysed to glycerol and three fatty acids.

Hydrolyzation: It means to cause or bring about hydrolysis or to cause a substance to split into component parts by the addition of water.

Hydrolyzed whey syrup: It is syrup made from whey by hydrolyzing the lactose into dextrose and galactose. These sugars are sufficiently soluble that the syrup can be made concentrated enough to be stable.

Hydrophile-lipophile balance (HLB value): The ratio of hydrophilic to hydrophobic groups on the molecules of an emulsifier.

Hydrophobins: These are the groups of small cysteine-rich, surface active proteins produced by filamentous fungi. They are known for their ability to form a hydrophobic or water-repellent coating on the surface of an object.

Hygrometer: An apparatus used for estimating the humidity of the atmosphere.

Hygroscopicity: It is the capacity of a product (e.g. packaging material) to react to the moisture content of the air by absorbing or releasing water vapor. Usually, a (dry) product is called hygroscopic if a small increase of water activity causes a considerable increase in water content.

I

Ice cream: It is frozen dairy product made by suitable blending and processing of cream and other milk products, together with sugar and flavor, with or without stabilizer or color and with the incorporation of air during the freezing process.

Ice milk: Ice milk is a general category of frozen dairy desserts that includes low-fat, reduced fat, light and non-fat ice cream. Ice milks resemble ice cream, except that they contain less fat than ice cream. Ice milks and ice cream contain stabilizers, emulsifiers and sugar.

Image analysis: Image analysis is the extraction of meaningful information from images; mainly from digital images by means of digital image processing techniques.

Imitation milks or imitation dairy products: Mixtures of non-dairy ingredients (other than milk, milk fat and non-fat milk solids) which are combined forming a product similar to milk, low-fat milk, or skim milk, or comparable analogs of other dairy products. Sodium caseinate, though derived from milk, is commonly termed a non-dairy ingredient and is often used as a source of protein in imitation milks.

Immobilized enzyme: An immobilized enzyme is an enzyme that is attached to an inert, insoluble material such as calcium alginate. This can provide increased resistance to changes in conditions such as pH or temperature. It also allows enzymes to be held in place throughout the reaction.

Immunoglobulins: Immunoglobulins are glycoprotein molecules that are produced by plasma cells in response to an immunogen and which function as antibodies. The immunoglobulins derive their name from the finding that they migrate with globular proteins when antibody-containing serum is placed in an electrical field.

In-bottle sterilization (sterilized milk): Sterilized milk usually is produced in two steps consisting of a continuous heating step at 120–140°C for several seconds and an in bottle sterilization at 110–120°C over 10–20 minutes.

In-container sterilization: In a batch in-container sterilization process, after preheating, the product is packaged. The second heating stage takes place in a sterilizer tower more than 10 m high. The tower is divided in several heating and cooling sections through which the filled bottles are transported.

Infant formula: It is a food which purports to be or is represented for special dietary use solely as a food for infants by reason of its simulation of human milk or its suitability as a complete or partial substitute for human milk. The most commonly used infant formulas contain purified cow's milk whey and casein as a protein source, a blend of vegetable oils as a fat source, lactose as a carbohydrate source, a vitamin-mineral mix, and other ingredients depending on the manufacturer.

Infusion: It is a variant of steam injection in which milk is sprayed into a steam flow or; in a steam-pressurized vessel, the product is sprayed into the top of the vessel, and while it falls, steam condenses on the product. This system is called the infusion or product-into-steam type.

Injection: In this system, steam, at a pressure higher than that of the product, is injected into the product stream through a suitable nozzle and condenses to give the required temperature. This system is called the injection or steam-into-product type.

Instant milk powder: Instant milk powder is produced by partially rehydrating the dried milk powder particles causing them to become sticky and agglomerate. The water is then removed by drying resulting in an increased amount of air incorporated between the powder particles.

Instantization: It refers to the process by which dried milk and milk products are made instant-soluble.

Interfacial energy: The free energy of the surfaces at an interface, resulting from differences in the tendencies of each phase to attract its own molecules; equal to the surface tension. It is also known as surface energy.

Interfacial tension: It is defined as the work which must be expended to increase the size of the interface between two adjacent phases which do not mix completely with one another. As a measure of work per unit area or force per wetted length, interfacial tension has the unit mN/m and is designated by the symbol γ (gamma).

Intermediate culture: An intermediate step in the manufacture of large volumes of bulk starter.

Intermediate moisture foods (IMF): These are semi-moist with about 25% (15-50%) moisture but with some of the water bound (which is unavailable to micro-organisms) by the addition of glycerol, sorbitol, salt, or organic acids, so preventing thc growth of micro-organisms.

International Dairy Federation: It is a non-profit private sector organization founded in 1903, representing the interests of various stakeholders in dairying at the international level. It represents the global dairy sector and ensures the best scientific expertise is used to support high quality milk and nutritious, safe and sustainable dairy products.

Invert sugar: It is a mixture of equal parts of glucose and fructose resulting from the hydrolysis of sucrose. It is found naturally in fruits and honey and produced artificially for use in the food industry.

Iodine value: Iodine value (I.V.) is a measure of the total number of double bonds present in fats and oils. It is generally expressed in terms of "number of grams of iodine that will react with the double bonds in 100 grams of fats or oils". High I.V. oil contains a greater number of double bonds than low I.V. oil. Edible oils with high iodine value are usually less stable and more susceptible to oxidation.

Ion exchange (whey): Ion exchange processing of whey is based on the ability of macromolecular resins to exchange their surface-bound ions for mobile ions of the same charge from the treated whey.

Ionic Strength: The ionic strength of a solution is a measure of the concentration of ions in that solution. Ionic compounds, when dissolved in water, dissociate into ions. The total electrolyte concentration in solution will affect important properties such as the dissociation or the solubility of different salts. One of the main characteristics of a solution with dissolved ions is the ionic strength.

Irradiated milk: It is the milk in which the vitamin D content has been increased by irradiation by ultra-violet rays.

Irradiation: The subjection of any material to the action of rays especially ultra-violet rays.

ISI: ISI mark is a certification mark for industrial products in India. The mark certifies that a product conforms to the Indian Standard, mentioned as IS: xxxx on top of the mark, developed by the Bureau of Indian Standards (BIS), the national standards body of India.

ISO 14000: The ISO 14000 family addresses various aspects of environmental management. It provides practical tools for companies and organizations looking to identify and control their environmental impact and constantly improve their environmental performance. ISO 14001:2004 and ISO 14004:2004 focus on environmental management systems.

ISO 22000: The ISO 22000 family of International Standards addresses food safety management. The ISO 22000 family contains a number of standards each focusing on different aspects of food safety management. **ISO**

22000:2005 (contains the overall guidelines for food safety management); ISO/TS 22004:2005 (contains guidelines for applying ISO 22000); **ISO 22005:2007** (focuses on traceability in the feed and food chain); ISO/TS **22002-1:2009** (contains specific prerequisites for food manufacturing); ISO/TS **22002-3:2011** (contains specific prerequisites for farming); **ISO/TS 22003:2007** (provides guidelines for audit and certification bodies)

ISO 31000:2009: It addresses risk management and provides principles, framework and a process for managing risk.

ISO 50001: ISO 50001 supports organizations in all sectors to use energy more efficiently, through the development of an energy management system.

ISO 9000: The ISO 9000 family addresses various aspects of quality management and contains some of ISO's best known standards. Standards in the ISO 9000 family include: **ISO 9001:2008** (sets out the requirements of a quality management system); **ISO 9000:2005** (covers the basic concepts and language); **ISO 9004:2009** (focuses on how to make a quality management system more efficient and effective); **ISO 19011:2011** (sets out guidance on internal and external audits of quality management systems).

ISO: ISO (International Organization for Standardization) is the world's largest developer of voluntary International Standards. International Standards give state of the art specifications for products, services and good practice, helping to make industry more efficient and effective. Developed through global consensus, they help to break down barriers to international trade.

Isomalt: It is a sugar substitute, a type of sugar alcohol made from beets and used primarily for its sugar-like physical properties.

J

Jaggery (gur): It is the most common form of sugar sweeteners in rural areas. It is considered as most nutritive sweetener. It is used by sweet-makers at domestic level in preparation of traditional sweets where a dark brown color is also desirable.

JECFA: The Joint FAO/WHO Expert Committee on Food Additives (JECFA) is an international scientific expert committee that is administered jointly by the Food and Agriculture Organization of the United Nations FAO and the World Health Organization WHO. Its work includes to evaluate the safety of food additives and the evaluation of contaminants, naturally occurring toxicants and residues of veterinary drugs in food

Junket: The soft coagulum or gel formed by the action of rennet on sweet milk.

Just-Hatmaker process: This is virtually the original method of drying milk by roller process. Two steam heated rollers at 250-270°F, practically touching, rotate inwards, the film of dried milk being scraped off by very accurately adjusted knives.

K

Kadhi: *Kadhi* is a well-known culinary food item prepared from *dahi* in many parts of India. Milk solid content in *kadhi* varies from 6-8 per cent and other solids range from 6-7 per cent. *Kadhi* exhibits a mildly acidic taste with characteristic cooked flavour. In the traditional method of *kadhi* preparation, 5-8 per cent of Bengal gram flour (*besan*) is added to stirred *dahi* or *buttermilk. Besan* acts as a thickening agent. In addition, boiled vegetables and fried balls of spiced Bengal gram batter (*pakora*) may be added at the end of preparation as filling materials. Salt and grated onion along with spices may also be used to garnish *kadhi*.

Kaelder milk: A Norwegian fermented milk of very good keeping quality in which lactic acid bacteria hydrolyze the lactose and torulae ferment the hexoses. The ripened culture contains about 0.5% alcohol and 2% or more lactic acid.

Kalajamun: It is similar to gulabjamun but darker in color. It is prepared either from khoa or chhana or mixture thereof. It has a hard crust outside and soft inside, with stuffing. It is also known as kalajam in northern India.

Kalakand: Kalakand is a milk sweet basically prepared from Danedar variety of khoa. Kalakand is characterized by large sized hard grains with less cohesive body. Contrary to burfi, kalakand has more distinct cooked flavour and brown colour.

Kashkaval cheese: Traditionally, in regions around the eastern part of the Mediterranean, pasta filata (i.e, stretched curd) cheeses are made. The curd is intensely heated and kneaded into loaves of small to medium size, and of several shapes. An almost unripened type is Mozzarella. Matured types are Provolone and Caciocavallo, known in the Balkans as Kashkaval.

Keeping quality: Keeping quality is an expression used to indicate the length of time milk or a milk product remains sweet and otherwise palatable and suitable for direct consumption. It may also be defined as the time elapsing from production or delivery to point of unfitness for use is called its keeping quality.

Kefir grains: These are the fermenting used for Kefir fermentation and contain Str. Lactis, Betabacterium caucasicum, Kefir bacilli and lactose fermenting yeasts.

Kefir: It is self-carbonated beverage containing 1% lactic acid and 1% alcohol originated from Caucasian Mountains.

Kefiran: Kefiran is the clear or pale yellow polysaccharide gel exuded by dairy kefir or water kefir grains.

Khandsari (khand): It is light to dark brown sugar having uniform crystals. It is also known as khand. It carries the natural sugarcane flavour as well as the aroma of natural caramel.

Khirmohan: It is a traditional Indian sweet dish that is made with fresh wheat flour, milk and butter. It is served warm as dessert.

Khoa powder: It is a product obtained by grating and drying the khoa. On reconstitution with water, this can be utilized directly for the preparation of burfi, milk-cake, kalakand and gulabjamun. Khoa powder packaged in tin containers under nitrogen gas can be stored for up to 10 months at 30°C.

Khoa/mawa: It refers to partially dehydrated whole milk product prepared by continuous heating of milk in karahi over a direct flame, while also constantly stirring and scraping by using khunti till it reaches to semi-solid consistency. According to Food Safety and Standard Regulations 2011, Khoya, by whatever variety of names it is sold such as Pindi, Danedar, Dhap, Mawa or Kava, means the product obtained from cow or buffalo or goat or sheep milk or milk solids or a combination thereof by rapid drying. The milk fat content shall not be less than 30 percent on dry weight basis of finished product. It may contain citric acid not more than 0.1 percent by weight. It shall be free from added starch, added sugar and added colouring matter.

Khunti: It is a long-handle flat-edged metal scraper used in khoa making to prevent caramelization.

Khurchan: This is a concentrated, sweetened whole milk product prepared by simmering milk without stirring it in a karahi, so as facilitate simultaneous formation of a thick creamy layer of skin on the surface of the milk, slow evaporation of water, after which sugar is added to the concentrated product.

Kishk: It is a traditionally dried product from the Middle East. It is made by combining low-fat yogurt (or buttermilk from churned, fermented milk) with salt and parboiled cracked wheat, known as burghol.

Kulfi: This is an indigenous ice cream frozen in small containers. Its mix is prepared from concentrated milk, sugar, malai, crushed nuts and flavor. The mix-in moulds is frozen in a large earthen vessel containing a mixture of ice and salt in the ratio of 1:1.

Kumiss: This is lactic acid-alcohol fermented milk originated in Russia. Culture consists of *L. acidophilus or bulgaricus.* The finished product may contain upto 2.5% alcohol.

Kunda: It is desiccated product, light brown to dark brown in color, prepared by the continuous heating of milk or high moisture khoa mixed with calculated amount of sugar. It has rich taste with pleasant caramel flavour.

L

LAB: The lactic acid bacteria (LAB) comprise a clade of Gram-positive, acid-tolerant, generally non-sporulating, non-respiring rod or cocci that are associated by their common metabolic and physiological characteristics. These bacteria, usually found in decomposing plants and lactic products, produce lactic acid as the major metabolic end-product of carbohydrate fermentation.

Laban (zeer): Sour milk produced during the hot season is transformed to a type of fermented milk called laban zeer, which has good quality. It has an acid and salty taste, with a characteristic strong yeasty aroma. The consistency of fresh laban zeer is syrupy but that of older samples is thick.

Labneh: Labneh is strained yogurt and has a cheese-like consistency. It makes a great alternative to sour or cream cheese.

Lactation: Lactation describes the secretion of milk from the mammary glands and the period of time that a mother lactates to feed her young.

Lactenins: Lactenin is a nitrogenous substance present in milk. It has antimicrobial properties. The spermicidal property of lactenin is important, as milk is used as a semen extender in artificial insemination of animals.

Lactic Acid: Lactic acid is a carboxylic acid with the chemical formula $C_3H_6O_3$. It has a hydroxyl group adjacent to the carboxyl group, making it an alpha hydroxy acid (AHA). Lactic acid is miscible with water and with ethanol, and is hygroscopic. It is produced from the fermentation of lactose in milk.

Lactic butter: Lactic butter is made from whole milk that has been fermented with lactic acid. It has a characteristic low moisture content and tangy flavor that some people find preferable to sweet cream butter.

Lactic cheese: These differ from the firmer rennet set cheeses in that they rely primarily on the activity of the bacteria converting lactose to lactic acid.

Lactitol: Lactitol is a sugar alcohol used as a replacement bulk sweetener for low calorie foods with approximately 40% of the sweetness of sugar. It is also used medically as a laxative.

Lactoferrin: Lactoferrin (formerly known as lactotransferrin) is a glycoprotein, and a member of a transferrin family, thus belonging to those proteins capable of binding and transferring ferric ions.

Lactometer: It is a special type of hydrometer used for measuring the specific gravity of milk covering the range 1.025-1.035.

Lactone: It is a cyclic ester which can be seen as the condensation product of an alcohol group -OH and a carboxylic acid group -COOH in the same molecule. It is characterized by a closed ring consisting of two or more carbon atoms and a single endocyclic oxygen atom, coupled with an adjacent ketone.

Lacto-peroxidase thiocyanate system: Lactoperoxidase together with its inorganic ion substrates, hydrogen peroxide, and oxidized products is known as the lacto-peroxidase system. The lactoperoxidase system plays an important role in the innate immune system by killing bacteria in milk and mucosal secretions hence augmentation of the lactoperoxidase system may have therapeutic applications.

Lactoperoxidase: It is a peroxidase enzyme secreted from mammary, salivary, and other mucosal glands that functions as a natural antibacterial agent. It belongs to heme peroxidase family of enzymes.

Lactose crystallization: Lactose crystallization occurs when the concentration of lactose exceeds its solubility. The physical properties of lactose crystals are dependent on the crystal type and can greatly influence their use in foods. Temperature affects the equilibrium ratio of the α- and β-lactose anomers. Lactose crystals formed at temperatures below 70°F (20°C) are mainly α-lactose crystals. The α-monohydrate lactose crystals are very hard and form, for example, when ice cream goes through numerous warming and freezing cycles. This results in an undesirable gritty, sandy texture in the ice cream.

Lactose intolerance: Some people suffer from lactose intolerance; they lack the lactase enzyme, hence they cannot digest lactose, or dairy products containing lactose.

Lactose: Lactose is a disaccharide (2 sugars) made up of glucose and galactose. It comprises 4.8 to 5.2% of milk, 52% of milk SNF, and 70% of whey solids. It is not as sweet as sucrose. When lactose is hydrolyzed by β-D-galactosidase, an enzyme that splits these monosaccharides, the result is increased sweetness, and depressed freezing point. It is also known as milk sugar and it is a type of reducing sugar.

Lactulose: It is a disaccharide made of fructose and galactose. It is a synthetic, non-digestible sugar used in the treatment of chronic constipation.

Ladder logic: Ladder logic is a programming language that represents a program by a graphical diagram based on the circuit diagrams of relay logic hardware. Ladder logic is used to develop software for programmable logic controllers (PLCs) used in industrial control applications.

Lalmohan: It is similar to gulabjamun and is made from chhana and lighter in color than gulabjamun. It is round in shape and has no stuffing inside.

Laminar flow system: Technological development for controlling the microflora in closed space (cabinet) is available which is known as laminar air flow system. In this system, air passes through high efficiency particulate air (HEPA) filters or ultra low penetration air (ULPA) filters.

Laminar flow: In laminar flow fluid moves slowly in layers in a pipe, without much mixing among the layers. Laminar flow typically occurs when the velocity is low and viscosity of the fluid is very high. Reynolds number lower than 2000 leads to laminar flow.

Langfil: It is a type of north Swedish cultured milk with a very rubbery consistency. In taste, it is very mild and soft.

Lanthionine: Lanthionine is a non-proteinogenic amino acid. As the monosulfide analog of cystine, lanthionine is composed of two alanine residues that are crosslinked on their β-carbon atoms by a thioether linkage. Despite its name, lanthionine does not contain lanthanum.

Lantibiotics: Lantibiotics are a class of peptide antibiotics that contain the characteristic polycyclic thioether amino acids lanthionine or methyllanthionine, as well as the unsaturated amino acids dehydroalanine and 2-aminoisobutyric acid.

Laplace pressure: The Laplace pressure is the pressure difference between the inside and the outside of a curved surface. The pressure difference is caused by the surface tension of the interface between liquid and gas.

Late blowing (cheese defect): It usually occurs during the ripening process of the cheese, most often after a minimum of three weeks storage and is primarily.

Latent heat: Latent heat is the energy released or absorbed by a body or a thermodynamic system during a constant-temperature process. For example phase transition such as the melting of ice or the boiling of water.

Leben: An Egyptian fermented milk of *L. bulgaricus* type. It is said sometimes to contain lactose-fermenting yeast which induces alcoholic fermentation.

Lecithin: Lecithin is a mixture of various phospholipids and the compositions depends on the origin of the lecithin. A major source of lecithin is soybean oil. Lecithin is admitted by the EU as a food additive and is designated by E number E322.

Leukocytes: These are the cells of the immune system that are involved in defending the body against both infectious disease and foreign invaders. These are also known as white blood cells.

Limburger cheese: Limburger is a semi-soft, washed rind cheese that originated in the historical Duchy of Limburg, now divided among three countries; Germany, Belgium and Netherlands. The cheese is popularly known for its stinky aroma which has been compared to foot odor.

Lipase: A lipase is an enzyme that catalyzes the hydrolysis of fats. Lipases are a subclass of the esterases. Lipases perform essential roles in the digestion, transport and processing of dietary lipids in most, if not all, living organisms.

Lipids: Lipids are a group of naturally occurring molecules that include fats, waxes, sterols, fat-soluble vitamins (such as vitamins A, D, E, and K), monoglycerides, diglycerides, triglycerides, phospholipids, and others. The main biological functions of lipids include storing energy, signalling, and acting as structural components of cell membranes.

Lipolysis: It refers to enzymatic hydrolysis of triacyl glycerols which causes free fatty acids to be formed and this may give the milk a soapy-rancid taste. Main lipolytic enzyme of milk is lipoprotein lipase. Optimum temperature of the enzyme is about 33□C and optimum pH is about 8.5.

Local area network (LAN): It is a computer network that interconnects computers within a limited area such as a home, school, computer laboratory, or office building using network media.

Locust bean gum: It is also known as Carob bean gum. It is a natural derivative from the seeds of the carob tree cultivated in the Mediterranean area. It is used for texturizing a wide range of food products.

Loss modulus: It is also known as viscous modulus. It measures the mechanical energy converted to heat through viscous frictional forces. It is represented by G”.

Loss tangent: It is ratio of loss modulus to the storage modulus in rheological measurements.

Low cholesterol: Maximum of 2 g saturated fat and a maximum of 20 mg cholesterol per serving are allowed.

Low-heat SMP: SMP which is given a pasteurization heat treatment and is used in the production of cheese, baby foods etc. where low degree of protein denaturation is required.

LTLT: It is a type of milk pasteurization which uses lower temperature for a longer time to kill pathogenic bacteria. In this type, milk is pasteurized at ~63°C for 30 minutes.

Lysine: Lysine is an α-amino acid. It is an essential amino acid for humans. Lysine is a base, as are arginine and histidine.

Lysozyme: It is an enzyme that catalyzes the breakdown of certain carbohydrates found in the cell walls of certain bacteria (*e.g.*, cocci). It is found in the secretions of the lacrimal glands of animals, nasal mucus, gastric secretions and egg white.

M

Maillard reaction: It refers to the interaction of amino compounds with sugar (casein with lactose) to form a brown color pigment called melanoidin. This reaction is known as Maillard reaction or Maillard type browning or non-enzymatic browning.

Makkhan: It refers to country/desi butter normally obtained by churning whole milk curd with crude indigenous devices known as 'Mathani'.

Malai: It refers to the creamy layer formed on the surface of milk, which has usually been heated slowly to boiling temperature and then left to cool undisturbed.

Malai-ka-baraf: This term is loosely applied to a variety of frozen products in which sweetened milk or malai may form the chief ingredient. They may be prepared from diluted milk thickened with fruit pulp, color and flavor.

Malted milk powder: It is the product obtained by mixing whole milk or partially skimmed milk or milk powder with malt extract and cereal grain flour in such a manner as to secure complete hydrolysis of the starchy material. It may contain cocoa powder and should be in the form of powder.

Malty flavor: This is not a common flavor but may be encountered in milk which is not properly cooled. Certain bacteria from improperly cleaned equipment, especially milking machines, may contaminate the milk and cause the objectionable malty flavor.

Manufactured products: Includes basically all dairy products except fluid milks: all cheeses, butter, ghee, curd, evaporated whole milk, condensed whole milk, condensed skim milk, whole milk powder, non-fat dry milk, ice cream, ice cream mix, frozen desserts, creams, and whey products.

Manufacturing automation protocol: It was a computer network standard released in 1982 for inter-connection of devices from multiple manufacturers. It was developed by General Motors to combat the proliferation of incompatible communications standards used by suppliers of automation products such as programmable controllers.

Margarine: It is a fatty solid butter substitute consisting of a blend of hydrogenated vegetable oils mixed with emulsifiers, vitamins, coloring matter, and other ingredients.

Market milk: It refers to packaged fluid milk that is sold to individuals usually for direct consumption.

Mastitis: Mastitis is the inflammation of the mammary gland and udder tissue, and is a major endemic disease of dairy cattle. Mastitis occurs when white blood cells (leukocytes), are released into the mammary gland, usually in response to an invasion of bacteria of the teat canal. Milk-secreting tissue and various ducts throughout the mammary gland are damaged due to toxins by the bacteria.

Maubois, Mocquot, Vassal (MMV) process: It is a cheese making process which uses ultrafiltration technology and is completely different from traditional cheese making process. The general idea behind MMV process is to concentrate all the milk proteins and simultaneously to remove excess water, some mineral ions and lactose by initial ultrafiltration carried out at the neutral pH of milk, prior to coagulation induced by enzyme and microorganism. Thus, ultrafiltration results in reduction or elimination of the need to separate the whey from the curd.

Maxwell material: It is a viscoelastic material having the properties both elasticity and viscosity. It is named after

James C. Maxwell who proposed the model in 1867. It is also known as Maxwell fluid.

MBRT: The methylene blue reduction test is based on the fact that the color imparted to milk by the addition of a dye such as methylene blue will disappear more or less quickly. The removal of the oxygen from milk and the formation of reducing substances during bacterial metabolism cause the color to disappear. The agencies responsible for the oxygen consumption are the bacteria. Though certain species of bacteria have considerably more influence than others, it is generally assumed that the greater the number of bacteria in milk, the quicker will the oxygen be consumed, and in turn the sooner will the color disappear. Thus, the time of reduction is taken as a measure of the number of organisms in milk.

Measurement error: Measured quantity value minus a reference quantity value.

Measurement: Process of experimentally obtaining one or more quantity values that can reasonably be attributed to a quantity.

Mechanical holes (cheese defect): It is a cheese defect in which holes of irregular shape are observed in cheese which are caused by trapped whey.

Mechanization: The use and development of machines to replace hand and labour.

Meleshin process: It is a process of butter making which is based on the inversion of high fat cream during scraped surface cooling using greater shear than Alfa type process with working applied for 150-200 seconds duration during the crystallization stage.

Meltdown: Meltdown is one of the important manifestations of ice cream structure. When ice cream is placed in an ambient environment to melt, two events occur; the melting of the ice and the collapse of the fat-stabilized foam structure. The melting of the ice is controlled by the outside temperature and the rate of heat transfer. However, even after the ice crystals melt, the ice cream does not "melt" or "collapse" until the fat-stabilized foam structure collapses. This is a function of the extent of fat destabilization/partial coalescence, which is generally controlled by the emulsifier concentration. The slow melt-down, slow serum drainage, good shape retention, and slower foam collapse are some of the desired important quality parameters of ice cream.

Membrane processing: Membrane processing is a technique that allows concentration and separation without the application of heat. In this technique, the particles are separated on the basis of their molecular size and shape with the use of pressure and specially designed semi-permeable membranes.

Mesophiles: These are the microorganism that grows best in moderate temperature, neither too hot nor too cold, typically between 20 and 45°C.

Metabolism: All the processes going in a living cell is known as metabolism. It is of two types: catabolism and anabolism. Catabolism refers to degradation type processes whereas anabolism refers to synthesis type reactions.

Metallic flavor: Metallic flavor is rough and puckery on the mouth and tongue. It is caused when milk comes into contact with corrodible metal, such as exposed copper on equipment or rusty milk cans or lids. It is very objectionable and may lead to further serious defects in certain dairy products such as butter.

Microfiltration: It is a method of removing contaminants in the size range 0.1 to 10.0 μm from fluids or gases. Microfiltration is achieved by passing liquids or gasses through a micro porous medium such as a membrane. There are two techniques used in microfiltration: dead-end filtration (for which microfiltration is widely used) and cross-flow filtration (which flow the fluid being filtered tangentially across the membrane).

Microsyneresis: It is a type of syneresis in which the exudation of liquid is from microscopic regions within a network.

MIDFT: Mansinhbhai Institute of Dairy & Food Technology is an academic institute established by Dudhsagar Dairy, Mehsana on its golden jubilee year 2010. It is named after the founder chairman of Dudhsagar Dairy Shri. Mansinhbhai Patel. It is the first institute established by a cooperative dairy.

Milk borne diseases (Transmissible): These diseases are transmitted from infected man to another via contaminated milk either directly by infected or carrier individuals, or indirectly through polluted water, insects, air & dust, milk utensils & equipment, rodents, flies, rations, etc. e.g. salmonellosis, bacillary dysentery, diphtheria etc.

Milk borne diseases (Zoonotic): These diseases are transmitted from infected animals to healthy individuals (human) by contaminated milk either directly from diseased or infected dairy cattle or indirectly via polluted water, flies, dust, contaminated milk equipment, etc. e.g. bovine tuberculosis, brucellosis, anthrax, Q-fever etc.

Milk can: It is a tall, cylindrical container with lid used for the transportation of milk.

Milk champagne: A popular name given to the effervescent alcoholic fermented milks such as kumiss.

Milk concentrate: Milk concentrated by evaporation, typically containing around 48% total solids.

Milk ices/milk lollies: These refer to the frozen product obtained from milk, skim milk or milk products with or without the addition of cane sugar, eggs, fruits, fruit juices, nuts, chocolates, edible flavors and permitted food colors. It may contain permitted stabilizers not exceeding 0.5% of the product. The mixture should be heat treated before freezing. The product should contain NMT 2.0% milk fat, NLT 3.5% proteins and NLT 20.0% total solids.

Milk plasma: Milk minus fat globule is called milk plasma or the liquid in which the fat globule float.

Milk powder functional properties: These are the properties of milk powder which make it useful for various food applications. Main functional properties of milk powder include water binding, browning, emulsification, whipping/foaming, gelling/thickening, dispersibility etc

Milk powder: Product obtained by removing water from pasteurized milk. It usually contains 3-4% moisture. Skimmed milk powder is often abbreviated "NFDM", which stands for non-fat dry milk.

Milk protein concentrate: Milk protein concentrate (MPC) is any type of concentrated milk product that contains 40–90% milk protein. MPC contains micellar casein, whey proteins and bioactive proteins in the same ratio found in milk. As the protein content of MPC increases, the lactose levels decrease. This high-protein low-lactose ratio makes MPC an excellent ingredient for protein-fortified beverages and foods and low-carbohydrate foods.

Milk separator: A milk separator is a centrifugal device that separates milk into cream and skimmed milk.

Milk serum: The liquid in which the micelles are dispersed or the milk minus fat globules and casein micelles.

Milking machine: An apparatus that attaches to a cow's or buffalo's teats and by vacuum draws the milk into a holding tank.

Milking parlour: A place separate from the barn where cows or buffalos are milked.

Milkoscan: It is an IR based instrument used to get fast and accurate milk testing analysis. It is generally pre-calibrated for milk and cream which enables it for on the spot analysis.

Milkotester: These are the equipments used for determining fat content of cream. These less sophisticated and less accurate than densitometer.

Milling: It refers to mechanical operation of cutting the blocks of cheddared curd into small pieces with the help of cheese mill.

Miscibility: The ability of two materials (especially liquids) to mix is known as miscibility.

Mishti doi or dahi: Mishti dahi is also known as Mishti doi, Payodhi and Lal dahi. It is a traditional sweetened fermented milk product popular in the eastern part of India, notably West Bengal, Bihar and Assam. It is prepared by lactic acid fermentation of sweetened milk. A good Mishti dahi has a characteristic brown colour, firm consistency, smooth texture and caramelized flavour.

MMPO (1992): In order to maintain the supply of desired quality of milk and to regulate the supply and distribution of milk and milk products in public interest and as per the de-licensing arrangement under the Government of India's licensing-free industrial policy, 1991, as per the Government of India notification number 350 dated 9-6-92, as also under the powers provided in section-3 of Essential Commodities Act 1955, the Central government issued the Milk and Milk Products Order-92, and it became effective on 9-6-1992.

Moisture content: Amount of water present in food products is known as its moisture content and is an important component of the food products as it directly controls the texture and microbial quality of the food. Moisture content is generally determined by heating samples at 130°C for at least 1 hour in a hot air oven.

Moisture: The moisture content of dairy products is of the greatest scientific and economic importance because free moisture controls bacterial growth and hence keeping quality.

Molding (Hooping): It refers to the curd being placed in hoops or moulds in which the cheese curd is pressed into its final shape.

Molds: These are the fungus that grows in the form of multicellular filaments called *hyphae*. Molds are large and taxonomically diverse number of fungal species where the growth of hyphae results in discoloration and a fuzzy appearance, especially on food.

Monoglyceride: Molecule consisting of one glycerol and one fatty acid (E471).

Monterey cheese: It is a semisoft cheese with a buttery-ivory color and a mild flavor similar to American Muenster. It is named after its creator, David Jacks, a cheese maker near Monterey, California. Jack cheese or Monterey cheese is made from pasteurized whole, partly skimmed or skimmed cow's milk. Unaged cheese is ripened for a week, has high moisture and good melting properties making it good for sandwiches and cooked dishes. Aged (dry) jack is a pale yellow color with a firm texture and a rich, sharp, slightly nutty flavor. Aged jack is often used as a grating cheese.

Mother culture: The culture prepared from the master culture at the dairy. The mother culture is prepared daily and is, as the name indicates, the origin of all cultures made at the dairy.

Mottled (cheese defect): Presence of irregular shaped spots or blotches in which portions of the cheese are light coloured and others are darker in colour, or an unevenness of colour due to mixing or combining the curd from two different vats.

Mousse: It is a prepared food that incorporates air bubbles to give it a light, airy and fluffy texture. It can range from light and fluffy to creamy and thick, depending on preparation techniques. These may be sweet or savory. Mousses are typically made with whipped egg whites or whipped cream, and generally flavored with chocolate or fruit puree.

Mozzarella cheese: It belongs to the cheese classified as 'pasta filata' which involves the principle of skilfully stretching the curd in hot water to get a smooth texture and grain in cheese. It is a soft, white un-ripened cheese which may be consumed shortly after manufacture. Its melting and stretching characteristics are highly appreciated in the manufacture of Pizza where it is a key ingredient.

Multiple effect evaporators: Two or more evaporator units can be run in sequence to produce a multiple effect evaporator. Each effect consist of a heat transfer surface, a vapour separator, as well as a vacuum source and a condenser. The vapours from the preceding effect are used as the heat source in the next effect.

Munster cheese: It is a strong tasting, soft cheese made mainly from milk. *Munster* is derived from the Alsace town of Munster, where, among Vosgian abbeys and monasteries, the cheese was conserved and matured in monks' cellars.

Musty flavor: This flavor is suggestive of musty or moldy hay. It may be absorbed directly by the milk but is more likely to come from feed or stagnant water consumed by the cow.

Mutarotation: Mutarotation is the change in specific rotation of a chiral compound due to epimerization or it may be defined as the change in the optical rotation because of the change in the equilibrium between two anomers, when the corresponding stereocenters interconvert. Cyclic sugars show mutarotation as α and β anomeric forms interconvert. The optical rotation of the solution depends on the optical rotation of each anomer and their ratio in the solution.

Mycotoxins: It is a toxic secondary metabolite produced by organisms of the fungi kingdom, commonly known as molds. These may also be defined as secondary metabolites produced by micro fungi that are capable of causing disease and death in humans and other animals.

Mysost cheese: Mysost cheese is Scandinavian in origin and traditionally made from the whey of cow milk. The color of this cheese is usually light to dark brown, depending on the amount of cream that has been added and the sugar caramelization levels.

N

Nanofiltration: Nanofiltration is a membrane filtration based method that uses nanometer sized cylindrical through-pores that pass through the membrane at a 90°. Nanofiltration membranes have pore sizes from 1-10 Angstrom, smaller than that used in microfiltration and ultrafiltration, but just larger than that in reverse osmosis.

National Cooperative Dairy Federation of India (NCDFI): It is the apex organisation for the cooperative dairy sector. Its members include federal dairy cooperatives of states and union territories. Primary objective of NCDFI is to facilitate the working of dairy cooperatives through coordination, networking and advocacy. It is based at Anand (Gujarat). Important activities of NCDFI includes; coordinating sale of milk and milk products of its members to the Ministry of Defence and other para military organizations, and marketing of frozen semen doses produced by Sabarmati Ashram Gaushala and Animal Breeding Centre.

National milk grid: National milk grid links milk producers throughout India with consumers in over 700 towns and cities, reducing seasonal and regional price variations while ensuring that the producers get fair market prices in a transparent manner on a regular basis.

NDDB: The National Dairy Development Board was created to promote finance and promote producer-owned and controlled organizations. Its programme and activities try to strengthen farmer co-operatives and support national authorities which favour the growth of such institutions. Fundamental to NDDB's efforts are cooperative principles and cooperative strategies.

Net protein utilization (NPU): It is the ratio of amino acid converted to proteins to the ratio of amino acids supplied. It may also be defined as the percentage of protein contained in the food which is retained by the body after the food has been eaten. It is used to describe the value or usefulness of certain proteins in a diet.

Neutralization: It refers to the chemical reaction in which an acid and a base react to form a salt.

New product development (NPD): It is a multi-step process which includes concept, ideation, design, test and release. Ideation, Design and Test steps are repeated over and over again, at varying depths of detail and on different subsystems, until the product design is complete.

Newton's laws of motion: Newton's laws of motion are a set of three physical laws that laid the foundation for classical mechanics. They describe the relationship between a physical body and the forces acting upon it, and its motion in response to said forces.

Newtonian fluids: The fluids that obey Newton's law of viscosity are known as Newtonian fluids.

Nisin: Nisin is a polycyclic antibacterial peptide with 34 amino acid residues used as a food preservative. It is produced by fermentation using the bacterium *Lactococcus lactis*. In the food industry, it is obtained from the culturing of *L. lactis* on natural substrates, such as milk or dextrose, and is not chemically synthesized.

NIZO process: During the early seventies, a research team at the Netherlands Institute for Dairy Research (NIZO) developed a continuous yoghurt making process based on the same two-stage fermentation i.e. the pre-fermentation (pH-stat) stage followed by the coagulum formation (plug-flow fermenter) stage.

Nodulation in yoghurt: Localized protein precipitation often referred to as 'nodulation' which is a major characteristic fault in stirred type yoghurt.

Non-Newtonian fluids: The fluids that do not obey Newton's law of viscosity are known as Newtonian fluids.

Non-protein nitrogen (NPN): Non-protein nitrogen (or NPN) is a term used in animal nutrition which refers collectively to components such as urea, biuret and ammonia, which are not proteins but can be converted into proteins by microbes in the ruminant stomach.

Non-starter LAB (NSLAB): Lactic acid bacteria associated with fermented dairy foods which include the Mesophilic *lactobacilli* and *pediococci sp.* found in ripening hard and semi-hard cheese and referred to as non-starter lactic acid bacteria.

Nuclear Magnetic Resonance (NMR): NMR spectroscopy is a technique used to determine the detailed, three-dimensional structure of molecules and, more broadly, to study the physical, chemical and biological properties of matter. It uses a strong magnet that interacts with the natural magnetic field in atomic nuclei.

O

Oatrim: It is a fat mimetic with a mouthfeel very similar to that of triglycerides. It is produced by partial enzymatic hydrolysis of oat flour and consists of 5% beta-glucan, soluble fiber, and amylodextrin.

Occlusion: The absorption of gases by solids e.g. of oxygen by milk powder.

Off-flavors: These are the undesirable flavor in the food which makes the product unacceptable for the consumers. Examples include soapy, astringent, rancid, bitter, papery, cowy etc.

Ohmic heating: Ohmic heating is an advanced thermal processing method wherein the food material, which serves as an electrical resistor, is heated by passing electricity through it. Electrical energy is dissipated into heat which results in rapid and uniform heating. Ohmic heating is also called electrical resistance heating, Joule heating or electro-heating and may be used for a variety of applications in the food industry.

Olestra: It is a fat-substitute consists of sucrose polyester structure of six to eight fatty acids bound together by a sucrose core. It is a synthetic fat composed of sucrose and edible oils. Olestra can be used as a fat substitute virtually anywhere because of its flexibility of chain length and saturation levels on the sucrose molecule.

Open texture (cheese): Cheese defect in which cheese contains irregularly shaped mechanical openings, caused by workmanship and not gas fermentation.

Operation Flood: Operation Flood was a project of the National Dairy Development Board (NDDB). It was the world's biggest dairy development program which made India, a milk-sufficient nation, the largest milk producer in the world, surpassing USA in 1998, with about 17 percent of global output in 2010–11.

Organic milk: It refers to the milk from livestock raised according to organic farming methods.

Organoleptic properties: Properties perceptible to the senses, especially taste and smell.

Osmotic pressure: It is the minimum pressure which needs to be applied to a solution to prevent the inward flow of water across a semi-permeable membrane.

Ostwald ripening: It is the phenomenon in which smaller particles in solutions dissolve and deposit on larger particles in order to reach a more thermodynamically stable state wherein the surface to area ratio is minimized.

Overrun (ice cream): It is the volume of ice cream obtained in excess of the volume of the mix. It is usually expressed as a percentage. This increased volume is composed mainly of the air incorporated during the freezing process.

P

Packaging: Packaging is defined as a co-ordinated system of preparing goods for transport, distribution, storage, retailing and end-use. It should perform one or more of the following functions: containment, protection, communication and convenience.

Paired comparison test: It is a two products (e.g. A & B) test, and the panellist's job is to compare these and identify whether the samples are similar of different. If different, which attribute, such as sweetness, acidity, hardness, colour etc. is responsible for this difference. This part of the test is called as directional difference test. The test can be further extended and the preference component of the panellist can be included. The paired comparison test is relatively easy to organize and implement. The two coded samples in order of AA, BB, AB, BA (4 pairs) are served simultaneously, and panellist has to decide if there is any difference or not. Paired comparison is typically used in comparing new and old processing techniques, change of ingredients in a product, preference testing at the consumer level, etc.

Paneer: It is a product obtained through heat/acid coagulation of casein component of standardized buffalo milk, entrapping through complex physic-chemical interactions almost all the fat, a part of denatured whey proteins and colloidal salts, as well as a part of soluble milk solids. Paneer is generally marble white in appearance, having a slightly spongy body, close-knit texture and possessing a sweetish-acidic-nutty flavor.

Pantua: It is a khoa and chhana based product sweet similar to gulabjamun. It is cylindrical in shape and has stuffings of nuts and saffron. The proportion of chhana is high compared to that of khoa.

Para-casein: The compound produced by the action of rennin on κ-casein (which liberates a glycoprotein), and that precipitates with calcium ion as the insoluble curd.

Paraffin wax: It is colourless or white, somewhat translucent, hard wax consisting of a mixture of solid straight-chain hydrocarbons ranging in melting point from about 48° to 66°C. Paraffin wax is obtained from petroleum by dewaxing light lubricating oil stocks. It is used in candles, wax paper, polishes, cosmetics, and electrical insulators. It assists in extracting perfumes from flowers, forms a base for medical ointments, and supplies a waterproof coating for wood. In wood and paper matches, it helps to ignite the matchstick by supplying an easily vaporized hydrocarbon fuel.

Paraffining: It refers to the operation of dipping the cheese for a few seconds in a bath of melted paraffin, whereby a thin coating of paraffin is applied to the surface of the cheese.

Parchment packing: Parchment is used for packing butter in boxes and casks and can be a source of mould spores leading to spoilage in butter. This can be minimized by treating the parchment for 24 hours with saturated brine.

Paring discs (Centripetal pumps): These are the devices present at the outlet of cream separators which converts the rotational energy of exiting milk and cream into linear kinetic energy and acts as stationary centripetal pump.

Parmesan cheese: It refers to few kinds of Italian extra-hard cheeses made of cow's milk. Parmesan is a part of Italian national cuisine and is usually grated. It is hard, sharp and dry.

Particle size distribution (PSD): The particle-size distribution (PSD) of a powder, or granular material, or particles dispersed in fluid, is a list of values or a mathematical function that defines the relative amount, typically by mass, of particles present according to size. It is important in understanding the physical and chemical properties of that material.

Pascal's law: A fluid under pressure exerts equal pressures on equal surface areas.

Passburg process: The German vacuum roller drier consists essentially of a steam-heated drum partly immersed in the milk which is drawn into the vacuum chamber. A knife removes the film of dried milk in the usual way, the capacity, being from 350-400 gal. per hour.

Pasta filata cheese: It is a cheese characterized by plasticity of the curd while it is being worked and molded e.g. provolone and mozzarella cheese.

Pasteurization: The process of heating milk or cream to a specific temperature for a specified time to destroy any potential harmful microorganisms and increase its keeping qualities.

Pasty body (cheese): Cheese defect in which cheese has a weak body and it becomes sticky and smeary when rubbed between the thumb and fingers.

Pathogens: Organisms capable of causing disease in animals or plants.

Payasam: It is a milk based delicacy popular in the southern parts of India. It resembles kheer of North India and has similar ritualistic connotations.

Pearson's square method: It is a method of standardization of milk. In this method, a square is drawn and desired fat content is placed in the centre of it. Place at the left hand corners of the square, the fat percentage of the materials to be mixed. Next, subtract the number in the centre from the larger number at the left hand side of the square and place the remainder at the diagonally opposite right hand corners. The number on the right hand side now represents the number of parts of each of the original materials that must be blended to have the desired fat content in resultant mix. The number at the upper right corner refers to the parts of material whose fat test was placed at the upper left corner and the number at the lower right corner refers to the parts of material whose fat test was placed at the lower left corner. If the numbers on the right are added, the sum obtained will represent the parts of the finished product.

Pectin: Pectin is a naturally-occurring thickening agent that is most often used by adding it to jams, jellies and similar products to help them gel and thicken. Pectin creates a thick, clear set when it gels. It is a complex carbohydrate found in and around the cell walls of plants, and helps to bind those cells together. All fruit has pectin in it, but the amount varies widely. Apples and oranges contain the most pectin, and the pectin from both fruits is used commercially to thicken many different types of products. Pectin generally needs high sugar content and some acid, such as citric acid, to activate, and some commercially available pectins include citric acid as an ingredient to help ensure that consumers get their desired result when working with their products.

Peda: Peda is a sweet prepared from pindi variety of khoa by the addition of sugar. Peda is characterized as a circular slightly flattened ball with low moisture content and white to creamy white in colour and smooth texture.

Penicillin: These are the antibiotics obtained from cultures of the molds *Penicillium notatum* or *P. chrysogenum*; interferes with cell wall synthesis in bacteria. All penicillins are β-lactam antibiotics. They are mainly bactericidal and are especially active against gram-positive organisms.

Pepsin: It is a proteolytic enzyme produced in the stomach in the inactive form pepsinogen, which, when activated by acid, splits proteins into peptones

Peptidases: Enzymes that perform proteolysis, that is, begins protein catabolism by hydrolysis of the peptide bonds that link amino acids together in the polypeptide chain forming the protein. They hydrolyse peptides into amino acids.

Per capita milk availability: It refers to the availability of milk per person per day. Per capita availability of milk in India in 2011-12 was 290 gram per day. Punjab and Haryana were having highest per capita availability of milk 945 and 720 gram per day, respectively.

Percolation threshold: It is a mathematical term related to percolation theory, which is the formation of long-range connectivity in random systems. Below the threshold a giant connected component does not exist; while above it, there exists a giant component of the order of system size. In engineering and coffee making, percolation represents the flow of fluids through porous media, but in the mathematics and physics worlds it generally refers to simplified lattice models of random systems or networks (graphs), and the nature of the connectivity in them.

Permeability: It is a measure of the ability of a porous material (or a membrane) to allow fluids to pass through it.

Permeation flux (membrane): The membrane permeation flux is defined as the volume flowing through the membrane per unit area per unit time.

Pesticides: These are the substances used to kill, repel, or control certain forms of plant or animal life that are considered to be pests. Pesticides include herbicides for destroying weeds and other unwanted vegetation, insecticides for controlling a wide variety of insects, fungicides used to prevent the growth of molds, disinfectants for preventing the spread of bacteria, and compounds used to control mice and rats.

PFA: Prevention of Food Adulteration (PFA) Act was promulgated by Parliament in 1954 to make provision for the prevention of adulteration of food, along with the Prevention of Food Adulteration Rules, 1955 which was incorporated in 1955 as an extension to the Act. Broadly, this Act covers food standards, general procedures for sampling, analysis of food, powers of authorized officers, nature of penalties and other parameters related to food. It deals with parameters relating to food additives, preservative, colouring matters, packing & labelling of foods, prohibition & regulations of sales etc. Like FPO, amendment in PFA rules are incorporated with the recommendation made by the Central Committee of Food Standards (CCFS) which has been setup by Central Government under the Ministry of Health and Family Welfare.

pH: Negative logarithm of hydronium ion concentration is known as pH. pH scale contains 1-14 scaling point in which 1-6.9 represents acidic pH, 7 represents neutral pH and 7.1-14 represents basic pH.

Phase inversion: Milk is an emulsion of fat in milk serum; butter an emulsion of milk serum in fat. The change of cream into butter is generally termed as phase inversion, i.e. the emulsion of fat in water becomes an emulsion of water in fat.

Phosphates: Monophosphates such as sodium and calcium phosphates and also di-, tri-, and poly-phosphates function as stabilizers in dairy products, though their predominant use in the dairy sector is as emulsifying salts for processed cheese and cheese preparations.

Phosphocaseinate: It is complex of calcium phosphate and casein manufactured from skimmed milk.

Phospholipids: These are a class of lipids and are a major component of all cell membranes as they can form lipid bilayers. Most phospholipids contain a diglyceride, a phosphate group, and a simple organic molecule such as choline; one exception to this rule is sphingomyelin, which is derived from sphingosine instead of glycerol.

Phosphotase: It is an enzyme that removes a phosphate group from its substrate by hydrolysing phosphoric acid mono-esters into a phosphate ion and a molecule with a free hydroxyl group.

Physical properties: These are the properties which are measurable and describe the physical state of the system. These can be measured without changing the composition of the material. Examples include appearance, texture, color, odor, melting point, boiling point, density etc.

Physical ripening: It serves to restructure the liquid fat after heating into a somewhat solid crystalline state, which is the basic requirement for a solid, plastified structure of the butter.

Phytosterol: It is a common sterol present in vegetable fat, analogous to and isomeric with cholesterol in animal fats and milk. The two sterols can be differentiated by crystalline form, melting points and optical rotations. The difference in melting points of their acetates forms the basis of a test for the adulteration of butter.

Pickering emulsion: It is a type of emulsion that is stabilized by solid particles (e.g. colloidal silica) which adsorb onto the interface between the two phases.

Picnometer: It is a standard vessel used in measuring the density or specific gravity of materials.

Pinny texture (cheese): Presence of numerous and very small gas holes in the cheese.

Pizza cheeses: These are the cheese varieties which are designed and manufactured for use specifically on pizza e.g. Mozzarella, Provolone, Parmesan etc.

Plastic cream: It is a type of cream containing as much fat as butter (80–83%) but as a dispersal of fat in water, while butter is water in fat. It is prepared by intense centrifugal treatment of cream. It is crumbly in texture and used for preparation of cream cheese and whipped cream.

Plastic deformation: When the stress is sufficient to permanently deform the material, it is called plastic deformation.

Plastic viscosity: It indicates the flow characteristics of a liquid when it is moving rapidly.

Plasticity: Capacity of fat to spread is known as its plasticity.

Platform tests: These include all those tests which are performed to check the quality of the incoming milk on the receiving platform, so as to make a quick decision regarding its acceptance or rejection. These are performed on each and every can or tanker of milk with the objective of detecting milk of inferior quality.

Plug flow: It is referred to as an idealised situation where there is no velocity profile across the tube. The fluid passes through the tube in same manner to a piston in a cylinder. There is a velocity distribution across the tube due to internal friction or viscosity of the fluid, which depends on whether the flow is streamline or turbulent.

Pneumatic conveying system: Pneumatic conveying system transfers powders, granules and other dry bulk materials through an enclosed pipeline. The motive force for this transfer comes from a combination of pressure differential and the flow of a gas, such as air or nitrogen. The system's basic elements include a fan or blower, feeder, conveying line, termination vessel and a dust collection system.

Poisson's ratio: It is the ratio of lateral strain to the axial strain. It is named after French physicist Simeon Denis Poisson.

Polydextrose: It is defined as randomly bonded polymer of glucose, sorbitol, and citric or phosphoric acid.

Polymerisation: Polymerization is a process of reacting monomer molecules together in a chemical reaction to form polymer chains or three-dimensional networks.

Polysorbates: Polysorbates are a class of emulsifiers used in some pharmaceuticals and food preparation. Polysorbates are oily liquids derived from PEG-ylated sorbitan (a derivative of sorbitol) esterified with fatty acids.

Portioning machine: It is used for the production of gulabjamun balls. The dough is fed in the hopper fixed on the top of the machine. The diaphragm type pump squeezes out required quantity of portions during each stroke.

Powder dispersibility: It is the ability of powder to separate into individual particles when dispersed in water with gentle mixing. Powders with high dispersibility typically exhibit good wettability and are agglomerated, with the absence of fine particles.

Powder recovery system: It is the system whose main function is to collect the overspray material and render it suitable for recycling and at the same time to remove the powder particles from the exhaust air stream before discharge into the atmosphere.

Power law: It is a special kind of mathematical relationship between two quantities. When the frequency of an event varies as a power of some attribute of that event (e.g. its size), the frequency is said to follow a power law.

Prebiotics: These are the non-digestible fiber compounds that pass undigested through the upper part of the gastrointestinal tract and stimulate the growth and/or activity of advantageous bacteria that colonize the large bowel by acting as substrate for them.

Preservatives: These are the substances which are added to food to enhance its shelf life such salt, sugar, acid etc.

Primary explosion: It is an explosion that takes place in the powder process plant itself.

Primost cheese: It is a Scandinavian cheese made from cow's milk which has a soft spreadable texture providing a semi-sweet taste. The sweetness is derived from caramelization of milk sugars. It is also known as mysost.

Principal component analysis: It is a statistical technique that uses an orthogonal transformation to convert a set of observations of possibly correlated variables into a set of values of linearly uncorrelated variables called principal components.

Probiotics: These are the live microorganisms (e.g. *Lactobacillus*, *Bifidobacteria* etc.) that are either the same as or similar to microorganisms found naturally in the human body and may be beneficial to health.

Processed cheese: It is a cheese product made from natural cheeses that may vary in degree of sharpness of flavor. Natural cheeses are shredded and heated to a molten mass. The molten mass of protein, water and oil is emulsified during heating with suitable emulsifying salts to produce a stable oil-in-water emulsion. Depending on the desired end use, the melted mixture is then reformed and packaged into blocks, slices, or into tubs or jars. Processed cheeses typically cost less than natural cheeses; they have longer shelf-life, and provide for unlimited variety of products.

Propionic acid: It is a food additive used as preservative in certain food products as it inhibits the growth of mold and some bacteria in food products at low level of addition.

Protein hydrolysates: Protein hydrolysates are the breakdown products of protein in the form of peptides and amino acids as a result of hydrolysis of protein either by acid, alkali or proteolytic enzymes.

Proteolysis: It refers to the enzymatic breakdown of proteins into peptides and amino acids which have distinct flavor such as bitter. Proteolysis also occurs during ripening of cheese and give characteristic flavor to cheese.

Protocooperation: It is an interaction between organisms of different species in which both organisms benefit, but neither is dependent on the relationship.

Provolone cheese: Provolone is an Italian cheese made from cow's milk whose origins lie in Southern Italy. It is classified into two forms: Provolone Dolce, which is aged for 2-3 months, has a pale yellow to white color and sweet taste. The other, Provolone Piccante is aged for more than four months and has a sharper taste.

Pure milk: The whole of the milk drawn at the time of milking from a healthy animal; but does not include milk mixed with any preservative or chemical or colouring matter of any kind.

Putrid butter: A most objectionable defect in butter due to the growth of gram-negative bacteria which attack protein putrefactively.

Q

QC (Quality control): Quality control (QC) is a procedure or set of procedures intended to ensure that a manufactured product or performed service adheres to a defined set of quality criteria or meets the requirements of the client or customer. QC is similar to, but not identical with, quality assurance (QA).

Quality assurance (QA): It involves a coherent system of activities that assures (guarantees) that the products meet a set of defined quality marks.

Quality: It is the sum of characteristics such as composition, grade and capability of products expressing their suitability for a defined utilization. It is the parameter for economic efficiency of products.

Quantity: Property of a phenomenon, body, or substance, where the property has a magnitude that can be expressed as a number and a reference.

Quarg cheese: It is unripened, concentrated lactic curd cheese made with low and high fat content. It is closely related to Baker's cheese. Due to its neutral flavor and smooth consistency, Quarg is a suitable ingredient in several dishes.

Quarg separator: It is a centrifugal machine used in quarg manufacture. It separates the clotted protein mass from whey by centrifugal force and discharged through small openings in the wall of separator bowl.

Queso Blanco cheese: Queso Blanco is a Mexican soft, unaged fresh cheese made out of pure cow's milk or a combination of cow and goat's milk. It is commonly used in the Iberian Peninsula and in several Latin American countries, including Mexico.

R

Rabadi: It is a fermented indigenous food of India. It is popular is North-Western semi-arid regions of India and can be prepared by mixing and fermenting flour of wheat, pearl millet, barley or maize with buttermilk in summer months at room temperature (40-45°C) for 4-6 h. The fermented product is boiled, salted to taste, cooled and consumed.

Rabri: Rabri is a concentrated, sweetened whole milk delicacy containing several layers of clotted cream. It is creamy white to caramelised in color, possesses viscous body containing several layers of clotted cream with a chewy texture and pleasant caramelised flavor.

Raita: It is made from *dahi* and served as an additional dish with meals. To prepare *raita, dahi* is lightly beaten, spiced and salted to taste.

Rajbhog: Rajbhog is a variety of rasogolla that is ballooned and so larger in size than rasogolla, being about 50-60 mm in diameter. It is blended with saffron that imparts an attractive golden color yellow shade to it.

Rancidity: It refers to the chemical decomposition of fats, oils and other lipids. Specifically, it is the hydrolysis and/or autoxidation of fats into short-chain aldehydes and ketones which are objectionable in taste and odor. It results in undesirable odors and flavors in food when occurs.

Random measurement error: Component of measurement error that in replicate measurements varies in an unpredictable manner.

Rasogolla: It is traditional Indian dairy product obtained from chhana stored and served in sugar syrup. It is snow white in color and possesses a spongy, chewy body and smooth texture. It is similar to ping-pong ball in shape and size.

Rasomalai: Rasomalai is rasogolla that has been flattened and blended with rabri. It is refrigerated and served chilled. It is a very delicate, chewy or spongy sweet that has delectable taste.

Raw milk: It is the milk at farm level that has not been treated in any way. Raw milk is not pasteurized, separated, standardized or homogenized.

Rebodying (cream): Keeping cream of 25% fat or more at 5°C for some time, then warming to about 30°C (e.g., 30 min) and cooling it again slowly produces a strong increase in its viscosity, and it may even gel. This process is called rebodying.

Recombinant chymosin: It is the chymosin enzyme produced by gene technology in which bovine gene is inserted into the DNA of microorganisms that then produce the enzyme.

Recombination: The process by which genetic material is broken and joined to other genetic material resulting in the production of new combination alleles.

Recombined butter: Milk fat and skim milk can be recombined to yield a butter-like product known as recombined butter. The manufacturing process used for and physical structure of recombined butter is practically identical to margarine. Recombined butter is firmer than natural butter of the same fat composition, due to absence of fat globules, so that all fat crystals can participate in networks and solid structures.

Recombined milk: Milk made by mixing skim milk powder, milk fat, water and possibly other components.

Reconstituted cream: Cream which is made by the re-emulsification of butter in milk with the aid of certain emulsifying agent such as glycerol monostearate.

Reconstituted milk: Product resulting from the mixing together and rehydration of a dried product of milk with water. For example, non-fat dry milk and water yields reconstituted skim milk. Adding cream or butter oil yield reconstituted whole or low-fat milk. Adding vegetable oil yields filled milk.

Recorder-controller (STLR): The STLR records the temperature of the milk and the time of day. It monitors, controls and records the position of the flow diversion device (FDD) and supplies power to the FDD during forward flow.

Recrating: It refers to a slatted wooden box or framework for packing or any completely enclosed boxlike packing.

Recrystallization: Recrystallization is a laboratory technique used to purify solids based on their different solubilities.

Reduced cholesterol: The product has an allowable maximum of 2 g saturated fat and a minimum 25% cholesterol reduction per serving.

Reference material: It is the Material or substance, one or more of whose property values are sufficiently homogenous and well established to be used for the calibration of an apparatus, the assessment of a measured method or assigning values to materials.

Reference measurement standard: Measurement standard designated for the calibration of other measurement standard for quantities of given kind in a given organization or at a given location.

Refractive index: Refractive index is a physical property that is related to the physical structure of the liquid sample through which light is passing. It is temperature dependent and usually given at 20°C.

Refrigerant: A refrigerant is a substance or mixture, usually a fluid, used in a heat pump and refrigeration cycle. In most cycles it undergoes phase transitions from a liquid to a gas and back again.

Refrigeration: It is a process in which work is done to move heat from one location to another. The work of heat transport is traditionally driven by mechanical work, but can also be driven by heat, magnetism, electricity, laser etc.

Regeneration: The method of using heat of a hot pasteurized milk, to warm up a cold, incoming raw milk. The cold milk also serves to cool the hot milk. In modern, efficient pasteurizing plant an efficiency of upto 90% can be achieved.

Regenerator: Heating and cooling energy can be saved by using a regenerator which utilizes the heat content of the pasteurized milk to warm the incoming cold milk.

Reichert-Meissl value: A method to determine the proportion of volatile water-soluble fatty acids present in fats (e.g. butter). It is expressed as the milliliters of 0.1 N alkali required to neutralize the acids obtained from five grams of fat by a specified method of saponification and distillation.

Relative humidity: It is defined as the ratio of the vapor pressure of air to its saturation vapor pressure. When vapor and temperature equilibrium are obtained, the water activity of the sample is equal to the relative humidity of air surrounding the sample in a sealed measurement chamber. Equilibrium relative humidity (ERH %) is obtained by multiplying water activity by 100.

Rennet activity: It is the number of millilitres of milk, which can be clotted by one millilitre of liquid rennet within 40 minutes at room temperature of 35°C.

Rennet casein: It is obtained by the enzymatic coagulation of skim milk. Rennet casein has characteristic bland, stable flavor and high nutritional value. Industrially, it is used for production of plastic. There is growing interest in the use of rennet casein as a food ingredient partly because of its suitability for use in making cheese analogues and partly because of its flavour stability. Enzymes used for production of rennet casein include chymosin and microbial rennet.

Rennet coagulation time (RCT): The time duration between rennet addition and onset of gelation in cheese making is known as rennet coagulation time or rennet clotting time.

Rennet substitutes: These are the milk clotting enzymes which can be used as a substitute for calf rennet in cheese making process. Rennet substitute includes papain, ficin, bromelain, microbial rennet and recombinant chymosin.

Rennet: Rennet is an extract from the fourth stomach of young ruminants, such as cows, goats, and sheep and is often used in cheese making. It contains a number of enzymes that are designed to help these animals digest their mother's milk, and when added to milk, it will cause the milk to coagulate, forming the curds and whey. The active enzyme in rennet is called chymosin or rennin.

Renneting: It is the process of addition of rennet enzyme to milk for curd formation during cheese manufacture.

Repeatability: The type of precision relating to the measurements made under repeatable conditions (i.e. same method, same material, same operator, same laboratory and narrow time period).

Reproducibility: The concept of precision relating to measurements made under reproducibility conditions (i.e. same method, different operator, different laboratories, different equipment and long time period).

Resazurin rennet test: This test is used to detect mastitic milk. The resazurin indicates bacterial cell activity in the milk by a reduction of color, while rennet clotting of the milk shows up any chemical abnormality.

Resolution: Smallest change in a quantity being measured that causes a perceptible change in the corresponding indication.

Retort: Retort is a cooking process that uses heat and pressure to cook food in its sealed package. Retort pouches conditions are quite demanding with temperatures typically ranging from 120°C to 130°C.

Reverse Osmosis (RO): A process in which a liquid such as milk or whey is pumped through a semi–permeable membrane under high pressure. The size of the pores in the membrane allows separation to occur at the molecular level. The liquid that passes through the membrane is called permeate, and the remainder is called retentate. Permeate from reverse osmosis is primarily pure water. Commonly used to reduce the volume of milk or whey prior to drying or further processing.

Rewet agglomeration: Rewet agglomeration involves dispersion of the fine powder in air, rewetting it by mixing it with small droplets of liquid to form loose agglomerates with point-to-point contact.

Reynolds number: The Reynolds number Re is a dimensionless number that gives a measure of the ratio of inertial forces to viscous forces. Whenever the Reynolds number is less than about 2,000, flow in a pipe is generally laminar, whereas, at values greater than 2,000, flow is usually turbulent. Actually, the transition between laminar and turbulent flow occurs not at a specific value of the Reynolds number but in a range usually beginning between 1,000 to 2,000 and extending upward to between 3,000 and 5,000.

Rheology: It is the study of flow (liquids) and deformation (semisolids and solids) of material under the effect of external applied force.

Ricotta cheese: It is an Italian whey cheese made from sheep (or cow, goat, or Italian water buffalo) milk whey left over from the production of cheese.

Rind formation (cheese): Rinds are the outside shell of cheese that forms during the cheese making process.

Rind treatment (cheese): The rind treatment generally involves providing the cheese with a layer of latex. It primarily prevents mold growth on the rind, but it can also protect against mechanical damage. Successive treatments with latex should be applied, and it should contain fungicides, e.g., natamycin.

Ripened cheese: These are the cheeses made by coagulating milk proteins with enzymes (rennet) and culture acids. These cheeses are then ripened (aged) by bacteria or mold. Cheddar, Swiss, Colby, brick and Parmesan are some examples of bacteria-ripened cheeses. Blue, Roquefort, Camembert and Brie are examples of mold-ripened cheeses.

Ripening agent: These are the substances which help in the ripening process of cheese making and give distinct flavor to cheese. This ripening process determines the features that define many different varieties of cheese such as taste, body and texture. Examples include lipolytic and proteolytic enzymes.

RO water: This is the water that is obtained after removing the majority of contaminants from it by pushing water under pressure through a semi-permeable membrane.

Ropiness: Ropiness is a defect in food caused by the growth of undesirable bacteria. It is usually characterized by stringiness, increased viscosity or sliminess. It is known to occur in dairy products, wine, beer and even meat.

Roquefort cheese: Roquefort cheese is a kind of blue cheese which comes from the south of France. It is made from sheep's milk.

S

Salt balance theory (heat stability): According to this theory, optimum heat stability is dependent on a certain ratio/balance of calcium plus magnesium-ions to phosphate plus citrate-ions (ratio for cow milk is 0.37 and for buffalo milk is 0.39). Any disturbance in the salt balance, due to the excess or deficiency of either group, accelerates heat coagulation.

Salting: It is the process of addition of common salt in the products such as table butter (2-3%) and cheese (1-4%). The salt affects durability, flavor and consistency of the product.

Sandesh: It is a chhana-based sweet popular in eastern India. It is known for its palatability, aroma and as a rich source of milk proteins, fat, sucrose and fat soluble vitamins. It has firm body a smooth texture. Three distinct varieties of sandesh are popular: soft grade (naram-pak), hard grade (kara-pak) and raw grade (kachagolla).

Sandiness: The sensation experienced in the mouth when a product such as ice cream or condensed milk contains small crystals of sugar or ice.

Sanitization: It is a cleaning process in which equipments and surfaces in the dairy plants are made free from dirt, germs etc. It can be done with or without use of sanitizers.

Sanitizers: These are the substances that are applied to non-living objects to destroy microorganisms that are living on the objects.

Saponification value: Number of milligrams of potassium hydroxide required to saponify 1g of fat under the conditions specified. The saponification value is a measure of the free and esterified acids present in fats and fatty acid esters.

Scalded cream: It is a thick cream made by indirectly heating full-cream cow's milk using steam or a water bath and then leaving it in shallow pans to cool slowly. During this time, the cream content rises to the surface and forms 'clots' or 'clouts'. Sometimes it is also known as clotted cream.

Scalding or cooking (cheese): This refers to the heating of curd cubes and it begins within 15 minutes of cooking.

Scaling up: Increasing a recipe for bulk production while keeping the ratio and proportions the same.

Scorched particles: These are the dark color charred particles generally originate from milk solids that have been held up somewhere in the process and as a result are burnt or overheated. Its main cause is deposits in the drying chamber, atomiser or air distributor. The particles that break off get mixed with the product to appear in the powder as dark specks. These may be present in concentrate if burning occurs in the evaporator.

SDS-PAGE: A very common method for separating proteins by electrophoresis uses a discontinuous polyacrylamide gel as a support medium and SDS to denature the proteins. The method is called SDS-PAGE.

Seamy (cheese defect): presence of white thread–like lines that form if the pieces of curd are not properly joined together.

Seamy color: It is an appearance or color defect in cheese which is portrayed when the cheese appears interlaced with light-colored lines around each original piece of curd. The seamy appearance defect usually results from physically altered curd surfaces caused by exuded or crystallized milk fat, uneven salting or moisture evaporation that probably occurred prior to curd pressing.

Secondary explosion: A secondary explosion is a subsequent ignition of the unburned dust coming out through the pressure relief system. The ignition source is the flames from the primary explosion.

Sedimentation: Sedimentation is the tendency for particles in suspension to settle out of the fluid in which they are entrained, and come to rest against a barrier. This is due to their motion through the fluid in response to the forces acting on them: these forces can be due to gravity, centrifugal acceleration or electromagnetism.

Seeding: It is the process of introduction of grains or crystals of a substance to induce nucleation e.g. grains in ghee and crystals in lactose manufacture.

Self-desludging separators: These are normally self-cleaning type separators in which the dirt is discharged periodically without interruption of the processes.

SEM: Scanning electron microscope (SEM) is a type of electron microscope that produces images of a sample by scanning it with a focused beam of electrons. The electrons interact with atoms in the sample, producing various signals that can be detected and that contain information about the sample's surface topography and composition.

Sensitivity: The difference in analyte concentration corresponding to the smallest difference in the response of the method that can be detected.

Sensor: A sensor is a converter that measures a physical quantity and converts it into a signal which can be read by an observer or by an instrument e.g., a mercury-in-glass thermometer converts the measured temperature into expansion and contraction of a liquid which can be read on a calibrated glass tube.

Sensory analysis: It is a scientific method used to "measure, analyze, and interpret human responses to products as perceived through their senses of touch, taste, sight, smell, or sound."

Separator slime: The visible dirt, cell debris etc., which is separated in skimming milk. It contains about 18-26% protein, 2.5-3.5% ash, 0.5-3.0% fat and 2-8% other organic matter. It is heavier than milk.

Sequestrants: These are the substances which combine with polyvalent metal ions to form a soluble metal complex, to improve the quality and stability of products.

Serum butter: Butter made from fat of whey or milk serum.

Serum proteins: The serum (whey) protein family consists of approximately 50% β-lactoglobulin, 20% α-lactalbumin, blood serum albumin, immunoglobulins, lactoferrin, transferrin, and many minor proteins and enzymes. Whey proteins do not contain phosphorus, by definition, but do contain a large amount of sulphur-containing amino acids.

Serum solids: The serum solids or milk solids-not-fat (MSNF) contain the lactose, caseins, whey proteins, and minerals (ash content) of the product from which they were derived.

Sewage treatment: The treatment of sewage to reduce its biological oxygen demand and to inactivate the pathogenic microorganisms present.

Shear modulus: It is the ratio of shearing stress to the shearing strain.

Shelf life: The period nominated by the manufacturer during which the product maintains its fitness for the intended purpose at a specific storage temperature and at other specified conditions. It may also refer to recommended period of time during which a food product may be stored and remain suitable for use.

Sherbet: It is a frozen dessert made from sweetened water with flavourings.

Short texture (cheese): There is no elasticity to the cheese plug removed by means of a grading trier, and the plug tends toward mealiness when rubbed between the thumb and fingers.

Shortening: Shortening is a semisolid fat used in food preparation, especially baked goods. Shortening contains 100% fat content as compared to 80% for butter and margarine.

Shrikhand: It is an Indian sweet dish made from strained yoghurt. For shrikhand preparation, yogurt is tied in a cotton cloth and left under pressure to drain. The strained yogurt, referred to as "Chakka" and sugar are mixed thoroughly. Cardamom, saffron, and any other flavors are then added and mixed. It is then left in the refrigerator for the sugar to dissolve and served chilled.

Single cell protein (SCP): Protein produced by microorganisms and primarily composed of microbial cells; sources of this protein include bacteria, fungi and algae.

Skim milk: Milk from which sufficient cream has been removed to reduce its milk fat content to not more than 0.2 %. It is rich in solid not fat (SNF) content and is used for standardization of milk, preparation of skim milk powder and coffee whitener. Skim milk contains protein, lactose, minerals and water–soluble vitamins and only half as many calories as whole milk.

Skuta: An alcoholic fermented whey made from sheep's milk in the Carpathians.

Skyr: An Icelandic soured or fermented milk.

Slide valve: A valve that slides back and forth over ports, especially one in the cylinder wall of a steam engine that permits the intake and outflow of steam to move the piston.

Slime layer: A capsular layer surrounding microbial cells composed of diffuse secretions that adheres loosely to the cell surface.

Slimy milk: It is a synonym used for ropy milk.

Slitty (cheese defect): Cheese contains narrow, elongated slits which are generally associated with a cheese that is gassy or yeasty. Slits in cheese having this characteristic may sometimes be referred to as fish eyes.

Smear ripened cheeses: These are the cheeses which are smear-ripened with solutions of bacteria or fungi, most commonly *Brevibacterium linens, Debaryomyces hansenii,* and/or *Geotrichum candidum*) which usually gives them a stronger flavour as the cheese matures.

Smell: It is a sensation produced in the olfactory centre of the brains by nerve impulses cause by the impact of molecules on the peripheral end of the olfactory nerve in the nose.

Smetana: It is a dairy product produced by souring heavy cream. It is from Central and Eastern Europe.

Smoky: It is a flavor defect in milk products (e.g. khoa) caused by use of smoky fire for boiling and dehydration of milk.

Soapy taste: A soapy taste in dairy products is usually due to lipolysis and alkaline reaction.

Sodium alginate: Sodium Alginate (E401) is extracted from brown seaweed. It is used as a stabilizer for ice cream, yogurt, cream, and cheese. It acts as a thickener and emulsifier for salad, pudding, jam, tomato juice, and canned products.

Sodium caseinate: Sodium caseinate is the most commonly used water soluble form of casein and is used in wide range of processed food products as a source of protein, and for their physicochemical, nutritional and functional properties. It is prepared from freshly precipitated acid casein curd or from dry acid casein by reaction with dilute solution of sodium hydroxide.

Soft water: The water that forms lather with soap easily is called soft water. It describes type of water that contains few or no minerals like calcium or magnesium ions.

Sogginess: It refers to a wet pasty or mushy condition either loaded with water or oil.

Sol: A colloid dissolved in water in colloidal solution is known as sol. Albumin and globulin in milk and whey are examples of a sol; heat may convert them into a gel form.

Solids–Non–Fat (SNF): The solids in milk other than milk fat; e.g., protein, lactose and minerals. The cells and bacteria are usually not of gravimetric importance.

Solubility index: It measures the amount of product sediment after the application of low centrifugal forces under specified conditions. High value of solubility index indicates that the product is less soluble.

Soluble soybean polysaccharide (SSP): It is a water soluble polysaccharide obtained from soybean and is mainly composed of galactose, arabinose and galacturonic acid. It can be used as dietary fiber fortification as well as stabilization, emulsification and adhesion.

Sorbic acid: It is 2, 4-hexadienoic acid. It is a natural, organic preservative frequently used to maintain the freshness of a variety of food products. It is a colourless solid, slightly soluble in water and sublimes readily.

Sorbitol: It is a sugar alcohol or polyhydric alcohol with half the sweetness of sugar. It can be obtained by reduction of glucose, converting the aldehyde group to a hydroxyl group. It is generally made from corn syrup.

Sorption isotherm: It describes the equilibrium of the sorption of a material at a surface at constant temperature. The water sorption isotherm is the relation between the equilibrium moisture content of a material and water activity, at a given temperature.

Sour cream: It is a dairy product obtained after fermentation of regular cream with some lactic acid bacteria.

Sour: Tart taste due to the presence of acid in foods; can be evidence of microbial spoilage or the desired result of controlled fermentation to produce a food product; caused by accumulation of acidic products of microbial metabolism.

Soymilk: Soy milk is made by soaking soybeans, grinding them with water. The fluid which results after straining is called soy milk.

Sparging: It involves bubbling a chemically inert gas such as nitrogen, argon or helium through a liquid. It can be used to remove dissolved gases (such as oxygen) from the liquid.

Specific gravity: It is the ratio of density of a substance to the density of a reference substance (water for liquid or gases).

Specific volume: It is defined as the number of cubic meters occupied by one kilogram of a particular substance or it is ratio of substance volume to its mass. The standard unit of specific volume is m^3/kg.

Sphingolipids: Sphingolipids are a class of lipids with a polar head group and two nonpolar tails. The core of a sphingolipid is an amino alcohol called sphingosine.

Sphingomyelins: Sphingomyelins are complex membrane lipids composed of phosphorylcholine, sphingosine, and an acylated group, such as a fatty acid.

Spray drying: It is a drying technique in which dry powder from a concentrated liquid or slurry is produced by rapidly drying with a hot air.

Spreadable butter: This is a type of butter which spreadable even at refrigerated temperature. It is made by mixing dairy butter with low melting point fat which makes it spreadable at low temperature. It do not need thawing at can be consumed as such.

St. Paulin cheese: Saint Paulin is a semi-soft monastery cheese with a moist, springy, orange rind made from pasteurized cow milk. The interior is white or light-yellow in colour. It has a tender, creamy consistency. Its flavour is milder, creamier and more buttery.

Stabilizers: These are the substances which absorbed water and are used to stabilize emulsified foods. These are also known as thickening agents. Examples of stabilizer include alginates, carrageenan, gelatine, guar gum etc.

Stainless steel: These are essentially alloys of iron, chromium, little carbon and some containing nickel.

Standardization: The process of adjustment of milk or cream fat content to a precisely specified or desired value.

Standardized milk: Milk which has been brought to a uniform milk fat and/or milk solids–not–fat content different from that of such milk at the ranch. Standardization of milk may be accomplished by the addition of skim milk, cream, or whole milk. Cream may also be removed to produce standardized milk.

Starter: A starter is a culture of one or more species or strains of lactic acid bacteria that is added to milk to ferment it. Sometimes the starter also contains non-lactic acid bacteria. Generally, starter is obtained via growth of lactic acid bacteria in milk at a suitable temperature.

Stassanisation: A method of heat-treatment in which milk is heated in very thin layers on the same principle as the HTST process.

Sterilized cream: It is a type of cream prepared via standardization, homogenization and sterilization (110-135°C/10-20 min). It may contain upto 0.2% stabilizing salts including calcium chloride and the sodium or potassium salts of carbonic, citric or orthophosphoric acids. It can be sold as "half and half" or coffee cream containing around 25% fat.

Sterilized milk: It is the name given to the homogenized milk which has been sterilized by exposing the sealed bottles to an atmosphere of steam usually between 220°F and 230°F for a period of 20-30 minutes.

Sticking process: There is a tendency of semi-dried material to stick to the hot metal surface inside the dryer during certain drying conditions. Powder particles may also stick to each other and form lumps during these conditions. This phenomenon is referred to as thermo plasticity or sticking.

Sticky (cheese defect): Cheese sticks to the palate and teeth when eaten.

Stilton cheese: It is a semi-soft blue-veined cheese made from pasteurized cow milk and was originated from England. It has creamy, crumbly and smooth texture and natural rind.

Stirred yoghurt: For manufacture of stirred yoghurt, the warm cultured milk mixture is incubated in a vat, cooled and then stirred for a creamy texture, often with fruit or other flavourings added. Stirred yogurt is often slightly thinner set-type yogurt.

Storage modulus: It is also known as elastic modulus. It measures the energy stored in viscoelastic solids. It is represented by 'G'.

Storage tanks: These are the containers which are used for the storage of milk and milk products for long duration of time. These may have facility of temperature control.

Storch & Segelcke rule: Storch & Segelcke (1874) demonstrated a rule related to clotting of milk in cheese making. According to this rule, the flocculation time is inversely proportional to the rennet concentration. Flocculation time is time duration between the addition of rennet and presence of first flocs as visually observed.

Stress relaxation tests: These are the tests carried out to determine stress-relaxation of a material, the specimen is deformed to a given amount and decrease in stress is recorded over prolonged period of exposure at constant elevated temperature. The stress-relaxation rate is the slope of the curve at any point.

Stretching (kneading): It is the defining operation in the manufacture of pasta filata pizza cheese. Stretching is typically performed using continuous single- or twin-screw mechanical mixers that contain hot water (50-55°C) coupled with steam injection systems.

Sublimation: The vaporization of a substance without passing through a liquid phase is known as sublimation. Drying in freeze dryers is based on the principle of sublimation.

Submicelles: Micelle is built of mixed-composition submicelles which are 12 to 15 nm in size and contains some 20 to 25 casein molecules.

Supercooling: The cooling of milk below the actual freezing point to obtain separation of ice.

Supersaturation: Supersaturation is a state of a solution that contains more of the dissolved material than could be dissolved by the solvent under normal circumstances or it may be defined as the solution containing an amount of a solute greater than that required for saturation.

Surface tension: It is the force acting per unit length at the surface of material (liquid). It is measured in N/m or dyne/cm. The surface activity of milk is related to proteins, fat, phospholipids and fresh fatty acids present in it. Homogenization and heat sterilization increase the surface tension of milk. Milk has a surface tension of 50 dyne/cm at 20°C.

Surfactants: These are the substances which when added to a liquid reduces its surface tension, thereby increasing its spreading and wetting properties. These are also known as surface active agents.

Surmjolk: It is an Icelandic cultured milk product prepared from whole milk or toned milk and may have added flavorings.

Suzme: It is a product obtained after straining yoghurt. It is a thick, creamy yogurt that can be spread on bread or mixed with cold water and salt to make a refreshing summer drink called suzmecal.

Svecia cheese: It is a semi-hard cow's milk cheese from Sweden. It has a creamy consistency, is light yellow in color, with a mildly acidic taste.

Svedberg unit: It is the unit in which sedimentation coefficient of a particle is generally expressed.

Sweet cream butter: Butter made from cream in which no ripening or development of acidity has taken place.

Sweet curdling: It is one of the common bacteriological defect to which milk is subject. In this defect, generally, no acid is produced; hence it is termed as sweet curdling.

Sweet whey: It is a by-product produced during manufacture of rennet types of hard cheese like cheddar or Swiss cheese.

Sweetened condensed milk: Sweetened condensed milk is milk that is concentrated by evaporation, to which sucrose is added to form an almost saturated sugar solution, after which it is packed.

Sweeteners: These are the substances added to food products for sweetening purpose. These are of two types: natural sweeteners (sucrose) and artificial sweeteners (saccharin, aspartame).

Swiss cheese: Swiss cheese is a generic name for several related varieties of cheese originally made in Switzerland. It is an American version of the Swiss Emmental and is known for being shiny, pale yellow. While it has a firmer texture than baby Swiss, its flavor is mild, sweet and nut-like. It has a savory, but not very sharp taste. Some types of Swiss cheese have a distinctive appearance. Some cheeses have holes known as 'eyes'. Swiss cheese without eyes is known as 'blind'. But in general, the larger the eyes in a Swiss cheese, the more pronounced its flavor.

Syneresis: It is a property of dairy products which occurs due to a rearrangement of the network, leading to an increase in the number of particle–particle junctions. The network then tends to shrink, thereby expelling interstitial liquid. In cheese making it is desirable whereas in yoghurt making it is undesirable.

Synthetic cream: A cream-like material prepared by emulsification of non-milk fats or hardened oils such as whale, ground nut etc. formerly with dried egg, lecithin, gelatine etc and now usually with CMC, GMS etc.

Synthetic milk: Synthetic milk is not milk but an artificial imitation of natural milk with a high degree of adulteration to increase the volume of milk and thereby the profit. Main components of synthetic milk are water, pulverized detergent or soap, sodium hydroxide, vegetable oil, salt and urea.

T

Tachometer: It is the instrument used to measure the speed of rotation.

Tallowiness: A taint in butter, milk powder and other dairy products caused by oxidation of unsaturated fatty acids, catalyzed by copper, light etc.

Taxonomy: Taxonomy is the science of naming, describing and classifying organisms and includes all plants, animals and microorganisms of the world.

Technology: The study of scientific knowledge applied to industrial process.

Tempering (casein): It is the step involved in the process of casein manufacture done after pressing, milling and drying of curd. Tempering means the holding of the casein for a period (24 h) to allow efficient cooling, hardening of the casein and evenness of moisture throughout the batch.

Tempering (cheese): Tempering of cheese refers to bringing it to the proper consistency. Cheese is brought from the curing or storage room to the blending room till it attains a temperature of 16-21°C in about 48 hours. This is called tempering of cheese and has the effect of softening it.

Tettemelk: It refers to the Scandinavian fermented milk usually made using previously prepared batch as starter. Tettemelk can be cultured with leaves of the plant *Pinguicula vulgaris* also known as butterwort.

Texture: Texture is the manner of structure, interrelation of parts, structural quality. It may also be described as how a food feels in ones mouth or way it chews. Texture can be described as 'rough', 'smooth', 'hard', 'soft', 'liquid', 'solid', 'lumpy', 'gritty' etc. For example smoothness in ice cream, hardness of cookie, softness of bread crumb.

Therapeutic properties: These are the properties of specific food which are related to treatment of disease or have curative action.

Thermal efficiency: It is a dimensionless performance measure of a thermal device such as combustion engine, boiler or a furnace. The thermal efficiency must be between 0% and 100%. Due to inefficiencies such as friction, heat loss, and other factors, thermal efficiencies are typically much less than 100%.

Thermal mechanical analysis: Branch of material science which studies the properties of materials as they change with temperature. This technique used in thermal analysis of materials.

Thermal properties: The properties of a material that determine how it reacts when it is subjected to excessive heat, or heat fluctuations over time.

Thermisation: Thermisation refers to the heat treatment of milk at sub-pasteurisation temperatures (typically 50–70°C for 5–30 s) on reception at the dairy to reduce the viable bacterial load in the milk and minimise changes in quality and processability prior to conversion into product.

Thermocompressors: They are used in wide range of processes to save energy. They compress low pressure steam to usable pressure levels and thereby reclaim the heat energy in the low pressure steam, which otherwise would have been wasted.

Thermophiles: These are the microorganisms with optimal growth temperatures between 60 and 108°C, isolated from a number of marine and terrestrial geothermally-heated habitats including shallow terrestrial hot springs, hydrothermal vent systems, sediment from volcanic islands, and deep sea hydrothermal vents.

Thermoplasticity: It refers to the property of material which make the material soft when heated and re-hardening occurs upon cooling without appreciable changes in the properties of the material.

Thirattupal: It is a traditional Indian sweet similar to *Peda*. It is popular in south India and is characterized by coarse granular texture. Slightly sour milk or small quantity of curd is added to hot milk during preparation of this sweet to get granular texture.

Three-stage drying: The three-stage drying involves transfer of the second drying stage into the base of the spray drying chamber and having the final drying and cooling conducted in the third stage located outside the drying chamber. It consists of a main drying chamber, static integrated bed chamber and vibro fluidized bed chamber.

Thromboelastography: It is a method of testing the efficiency of blood coagulation. It is a viscoelastic hemostatic assay that measures the global viscoelastic properties of whole blood clot formation under low shear stress

Tilsit cheese: It is a light yellow semi-hard smear-ripened cheese. Tilsiter has a medium-firm texture with irregular holes or cracks. Commercially produced Tilsiter is made from pasteurized cow's milk and has a dark yellow rind.

Timing pump: The timing pump draws product through the raw regenerator and pushes milk under pressure through pasteurized regenerator. It governs the rate of flow through the holding tube.

Titratable acidity: It is a measure of titratable hydrogen ions. Includes H+ ions free in solution and those associated with acids and proteins.

Tocopherol: These are the class of compounds which have vitamin E activity. These are fat-soluble antioxidants but also seem to have many other functions in the body. The main sources of tocopherols are oils.

Tofu: It is also known as "soya paneer". It is made from soybeans, water and a coagulant, or curdling agent. It is high in protein and calcium and well known for its ability to absorb new flavors through spices and marinades.

Toned milk: It refers to milk obtained by the addition of water and skim milk powder to whole milk. In India, as per PFA Rules (1976), the toned milk should contain a minimum of 3.0 per cent fat and 8.5 per cent solids not fat. It is also known as single toned milk.

Toppings: These are the edible food materials that are applied on the top of the food for garnishing and decoration purpose e.g. pizza toppings.

Torba: It is traditional fermented product made from cow or eve or mixed milk throughout Turkey. It is concentrated or strained yoghurt obtained when whey is allowed to drain from fresh yoghurt.

Toughness: It is a texture related term which describes the ability of food material to absorb energy and plastically deform without fracturing.

Traceability: Traceability means the ability to track any food, feed, food-producing animal or substance that will be used for consumption, through all stages of production, processing and distribution. It can also be defined as the ability to verify the history, location or application of a food or food related item by means of documented recorded identification.

Trans fat: It is the common name for unsaturated fat with *trans*-isomer fatty acid(s). Trans fats may be monounsaturated or polyunsaturated but never saturated.

Trans fatty acids: Unsaturated fatty acid in which one or more double bonds are in the trans configuration i.e. in opposition to the cis fatty acids found in most fats and oils. Found in body and milk fat from ruminants (e.g. butter) and is produced by partial hydrogenation of fats and oils. These are considered as harmful for human health.

Transmission electron microscope (TEM): It is a conventional electron microscope which produces an image of a cross-sectional slice of a specimen all points of which are illuminated by the electron beam at the same time.

Triangle test: It is a differential sensory analysis technique. A triangle test is a type of difference test to determine if there is a sensory difference between two products.

Turbulent flow: Turbulent fluid flow is that in which the velocity at a given point varies erratically in magnitude and direction. Reynolds number higher than 4000 leads to turbulent flow.

U

UHT: A UHT product produced by UHT treatment is packaged in a sterile container under aseptic conditions. Typical temperature-time combination for UHT milk is 132°C for not less than 1 sec and for UHT cream 132°C for not less than 2 sec. Typically, UHT milk is packaged in aseptic containers or pouches do not require refrigeration.

Ultracentrifuge: A high-speed centrifuge which generates centrifugal force upto several hundred thousand times to force of gravity. It is used for sedimentation and molecular weight estimation of macromolecules.

Ultrafiltration (UF): It is a type of membrane filtration technology. When applied to skim milk, the albumin protein may pass into permeate but the casein proteins are captured in the retentate. It is generally used as a pre-treatment before cheese making.

Unripened cheese: These are the cheeses are made by coagulating milk proteins (casein) with acid. Examples include soft cheeses like cream cheese, cottage cheese. Ripening or ageing step is not involved in manufacturing procedure of these cheeses. With no rind and a soft and smooth texture, they are high in moisture, generally lower in fat. They have a shorter shelf life.

V

Vacreator/ Vacreation: It is the equipment in which steam is intimately mixed with cream and the condensed vapor plus volatiles are removed by flash evaporation under vacuum. The process is known as Vacreation.

Vacuum evaporation: The process of concentrating a liquid product using vacuum is known as vacuum evaporation.

Vacuum evaporator: These are the equipments used to concentrate milk before drying using vacuum.

Vacuum producing equipments: These are the equipments used to produce and maintain vacuum in the plant by removing the air and non-condensable gas. Water ring pump and steam jet ejectors are used generally used as vacuum producing equipment.

Valio system: This is system used for hydrolysis of sweet wheys utilizing the enzyme from *Aspergillus niger*. The system operates at a pH of about 4.0 and a temperature of 40°C.

Vapor-compression evaporation: Vapor-compression evaporation is the evaporation method by which a blower, compressor or jet ejector is used to compress, and thus, increase the pressure of the vapor produced. If compression is performed by a mechanically driven compressor or blower, this evaporation process is usually referred to as Mechanical Vapor Recompression (MVR).

Vapour-liquid separator: It is a device used in several industrial applications to separate a vapor-liquid mixture.

VDI 3673: VDI 3673 is the method used to measure pressure relief area in powder plants.

Vegetable rennet: It is a natural enzyme that coagulates milk and separates the curds and whey when you are making cheese. Commercial vegetable rennet is often extracted from a mold called *mucor miehei*. Vegetable rennet is used by vegetarian cheese makers to avoid adding rennet made from animal parts.

Velocity gradient: The difference in velocity between adjacent layers of the fluid is known as a velocity gradient and is given by v/x, where v is the velocity difference and x is the distance between the layers.

Vent ducts: These are the ducts that are connected to the pressure relief device and lead into the atmosphere at a safe place.

Vent ratio methods: The vent ratio method is generally used to measure pressure relief area. It is based on a maximum velocity of pressure increase measured in Hartmann's apparatus.

Ventex valve: At an explosion, it is the valve that is closed automatically by the explosion pressure. Ventex valve can also be detector controlled in which the detector will ensure closing of the valve by means of compressed air.

Viili: It is a Finnish product and is made of pasteurized non-homogenized milk. A polysaccharide-producing starter, comparable to that used for langfil, is employed. The milk is incubated at 18°C to 19°C for 18 to 20 h. A mold, *Geotrichum candidum*, is also added to the milk. A cream layer is formed, on which the mold creates a velvety layer and causes some hydrolysis of fat.

Viscoelastic material: A viscoelastic material exhibits viscous as well as elastic behavior.

Viscosity: Viscosity is a measure of a fluid's resistance to flow. It describes the internal friction of a moving fluid. A fluid with large viscosity resists motion because its molecular makeup gives it a lot of internal friction. A fluid with low viscosity flows easily because its molecular makeup results in very little friction when it is in motion.

Viscous material: In a purely viscous material all energy added is dissipated into heat. When a force is applied to a viscous material the material does not stretch, it flows like a liquid. When the force is removed it does not return to its original shape because the force has been 'used up' in the fluid flow.

Vitaminized milk: This is liquid or powdered milk which are artificially fortified with vitamins.

Voluminosity (micelles): It is the surface and structural property of casein micelles which leads to the increased viscosity.

W

Wafers: These are crisp, often sweet, very thin, flat, and dry biscuits often used to decorate ice cream. These can also be used in making cream flavoring sandwiched cookies.

Water activity: It is the ratio of the vapor pressure of water in a material (p) to the vapor pressure of pure water (p_o) at the same temperature. Water activity is a measurement of the availability of water for biological reactions. It determines the ability of micro-organisms to grow. If water activity decreases, micro-organisms with the ability to grow will also decrease.

Water ices: These are the sweetened frozen desserts prepared from fruit concentrates or fruit juices or other natural or artificial food flavourings. These are similar to sorbets.

Water softening: It involves the removal of calcium, magnesium, and certain other metal cations in hard water. The resulting soft water is more compatible with soap and extends the life of pipelines used in dairy plants. Water softening is generally done using lime softening or ion-exchange resins.

Weak (cheese defect): Cheese requires little pressure to crush and is soft but not necessarily sticky like a pasty cheese.

Weaving flow: It is the counter-current flow of cream in the vacuum chambers of the Vacreator in which the cleanest cream meets the cleanest steam.

Weeping: Descriptive term referring to Swiss-type cheeses whose eyes glisten with bits of moisture. This is caused by the release of moisture by proteins as they are broken down during ripening. Weeping often indicates that a cheese has achieved peak ripeness and will exhibit full flavor. It can also be caused by storing cheese at too warm a temperature.

Westphal balance: This balance is used to measure the specific gravity (or density) of liquids.

Wet cleaning: It involves soaking, spray washing, flotation washing and ultrasonic cleaning methods. In wet cleaning method, water and detergent or sanitizers are generally used for removing adhering dirt. Wet cleaning often requires dewatering step to provide material or equipment suitable for processing and storage.

Wet scrubbers: These are used in powder plants. The basic principle of wet scrubber is to dissolve any dust powder left in the air stream into either water or the feed stream by spraying the wash stream through the air. It also recovers heat from the existing air and evaporates some of the water in the feed stream.

Wet solids: Condensed milk, skim milk, or whey may be referred to as wet solids, to distinguish from dry solids in the form of non-fat dry milk or dried whey powder.

Wettability: This defines the potential for a powder to wet and absorb water at a given temperature. In wetting, the voids within a powder are replaced by water.

Wetting agent: These are the substances which lower the surface tension of liquid and allow it to spread more rapidly e.g. propylene glycol, sodium lauryl sulfate etc.

Wetting rate: It is the rate at which dairy (milk) powders disperses and dissolves in the solvent.

Whey cheese: It refers to the class of cheese which are prepared from whey e.g. Ricotta, Mysost etc.

Whey cream: It is the fat rich portion separated from the whey using centrifugal separator. The whey cream may be utilized as such or churned to produce butter.

Whey powder: It is a product obtained by condensing (20-30%TS) and dried (roller or spray dried) the whey obtained from cheese industry.

Whey protein concentrates: Whey protein concentrate most common form of whey protein, a by-product of cheese production. Concentrates typically have a low (but still significant) level of fat and cholesterol but, in general, have higher levels of bioactive compounds as well as carbohydrates in the form of lactose. They are only 29%–89% protein by weight.

Whey protein isolates: These are the whey products which contain more than 90% protein on dry matter basis.

Whey protein nitrogen index (WPN): WPN index is a measure of the heat treatment applied to the milk during processing to milk powder. It is the basis of the heat classifications of milk powders: High heat powder (WPN=<1.5), medium heat powder (>1.5 -<6.0), low heat powder (>6.0). WPN is expressed as milligrams (mg) undenaturated whey protein nitrogen per gram of non-fat milk powder with a moisture content of 3.16%.

Whey proteins: These are the collection of globular protein from whey which is the by-product of cheese industry.

Whey: It is the greenish translucent liquid which is obtained during the manufacture of cheese, casein, chhana, paneer, chakka and co-precipitates. It contains about 93.3 percent water and 6.5 percent lactose, protein, minerals, enzymes, water–soluble vitamins and about 0.2% fat.

Wheyvit: It is a fermented whey beverage prepared by fermentation of whey with culture of *Saccharomyces cerevisiae*. It may be carbonated or non-carbonated.

Whipping cream: When cream is subjected to mechanical beating, air is incorporated and fat globules collect at the air-water interface and the milk fat globule membrane becomes disrupted. This is known as whipping cream and it contains about 30-40% fat.

Whipping: It refers to beating of cream to produce froth or foam (emulsion of air in liquid).

White butter: It is a fat concentrate obtained by churning cream, gathering the fat into a compact mass and then working it. It contains no added salt.

Whole milk powder: It is a dried milk product obtained from cow or buffalo milk by removal of water. It should contain maximum 5% moisture, minimum 26% fat and maximum 1.2% acidity. The maximum solubility index should be 15 for roller dried and 2 for spray dried product.

Working: Working is achieved by forcing the butter grains through a series of perforated plates between which are mixing vanes. The degree of working is controlled by a throttle plate which can be adjusted to vary the back pressure on the butter.

XYZ

Xanthan gum: It is a microbial gum used as a stabilizer and thickener in various dairy and food products. It is obtained from the bacteria *Xanthomonas campestris*. Its E-number is E-415.

Yield: The amount of product obtained from a fixed quantity of input material. Specific to pricing formulas used in the dairy industry, yield refers to how much butter, powder or cheese can be obtained from 100 kg of milk.

Ymer: It is fermented milk beverage prepared by souring of skim milk by ordinary starter and concentration effected by separation of the curd. After addition of cream, the milk is homogenized and filled into bottles at around 3°C.

Yogurt: It is a fermented milk product generally prepared from low-fat milk, or skim milk. Fruit, flavors and sugars may be added. Generally, yogurts are high in protein and low in calories. Sometimes referred to as refrigerated yogurt to distinguish from frozen yogurt, an ice cream–like product.

Zero adhesion technology (ZAT): It has been implemented in ice cream manufacturing machinery to produce novel ice cream presentations. Application ZAT permits the finished ice cream to release from their mold and be ready for packaging and any subsequent hardening.

Zeta potential: It is the potential difference between dispersion medium and the stationery layer of the fluid attached to dispersed particles. It represents the stability of colloidal dispersions.

Zymogen: The zymogen is called prochymosin and, by acid treatment (activation), it is converted to the active enzyme, i.e. chymosin.

α-lactalbumin: These proteins contain eight cysteine groups, all involved in internal disulfide bonds, and four tryptophan residues. alpha-lactalbumin (molecular weight 14000) has a highly ordered secondary structure, and a compact, spherical tertiary structure. Thermal denaturation and pH <4.0 results in the release of bound calcium.

β-lactoglobulin: This group, including eight genetic variants, comprises approximately half the total whey proteins. β-lactoglobulin (molecular weight 18000) has two internal disulfide bonds and one free thiol group. The conformation includes considerable secondary structure and exists naturally as a noncovalent linked dimer. At the isoelectric point (pH 3.5 to 5.2), the dimers are further associated to octamers but at pH below 3.4, they are dissociated to monomers.